A Text Book for F.Y.B.Sc. (Computer Science)
Semester - II
Mathematics Paper - I : MTC-121 (Credit 2)
Choice Based Credit System (CBCS)(2019 Pattern)

LINEAR ALGEBRA

Dr. Kalyanrao Takale
M.Sc., B.Ed., Ph.D.
RNC Arts, JDB Commerce and NSC Science College
Nashik Road, Nashik

Dr. Shrikisan Gaikwad
M.Sc., B.Ed., M.Phil., Ph.D.
New Arts, Commerce and Science College
Ahmednagar

Dr. Mrs. Nivedita Mahajan
M.Sc., M.Phil., Ph.D.
Modern College of Arts, Science and Commerce
Shivajinagar, Pune

Dr. Amjad Shaikh
M.Sc., Ph.D.
AKI's, Poona College of Arts, Science and Commerce
Pune

Prof. Mrs. Shamal Deshmukh
M.Sc., M.Phil.
Modern College of Arts, Science and Commerce
Ganeshkhind, Pune

Prof. S. R. Patil
M.Sc.
Ex. HOD., S.M. Joshi College
Hadapsar, Pune

N5064

LINEAR ALGEBRA   **ISBN 978-93-89825-00-8**

First Edition	:	**December 2019**
©	:	**Authors**

Published By :
NIRALI PRAKASHAN
Abhyudaya Pragati, 1312, Shivaji Nagar
Off J.M. Road, PUNE – 411005
Tel - (020) 25512336/37/39, Fax - (020) 25511379
Email : niralipune@pragationline.com

➢ DISTRIBUTION CENTRES

PUNE

Nirali Prakashan : 119, Budhwar Peth, Jogeshwari Mandir Lane, Pune 411002, Maharashtra
(For orders within Pune)   Tel : (020) 2445 2044, Mobile : 9657703145
Email : niralilocal@pragationline.com

Nirali Prakashan : S. No. 28/27, Dhayari, Near Asian College Pune 411041
(For orders outside Pune)   Tel : (020) 24690204; Mobile : 9657703143
Email : bookorder@pragationline.com

MUMBAI

Nirali Prakashan : 385, S.V.P. Road, Rasdhara Co-op. Hsg. Society Ltd.,
Girgaum, Mumbai 400004, Maharashtra; Mobile : 9320129587
Tel : (022) 2385 6339 / 2386 9976, Fax : (022) 2386 9976
Email : niralimumbai@pragationline.com

➢ DISTRIBUTION BRANCHES

JALGAON

Nirali Prakashan : 34, V. V. Golani Market, Navi Peth, Jalgaon 425001, Maharashtra,
Tel : (0257) 222 0395, Mob : 94234 91860; Email : niralijalgaon@pragationline.com

KOLHAPUR

Nirali Prakashan : New Mahadvar Road, Kedar Plaza, 1st Floor Opp. IDBI Bank, Kolhapur 416 012
Maharashtra. Mob : 9850046155; Email : niralikolhapur@pragationline.com

NAGPUR

Nirali Prakashan : Above Maratha Mandir, Shop No. 3, First Floor,
Rani Jhanshi Square, Sitabuldi, Nagpur 440012, Maharashtra
Tel : (0712) 254 7129; Email : niralinagpur@pragationline.com

DELHI

Nirali Prakashan : 4593/15, Basement, Agarwal Lane, Ansari Road, Daryaganj
Near Times of India Building, New Delhi 110002 Mob : 08505972553
Email : niralidelhi@pragationline.com

BENGALURU

Nirali Prakashan : Maitri Ground Floor, Jaya Apartments, No. 99, 6th Cross, 6th Main,
Malleswaram, Bengaluru 560003, Karnataka; Mob : 9449043034
Email: niralibangalore@pragationline.com

Other Branches : Gujarat, Kolkata Hyderabad, Chennai

niralipune@pragationline.com | www.pragationline.com
Also find us on f www.facebook.com/niralibooks

Preface

This book is based on a course Linear Algebra. We have written this book as per the revised syllabus of F.Y. B.Sc.(Computer Science) Mathematics, revised by Savitribai Phule Pune University, Pune, implemented from June 2019. Linear Algebra is the most useful subject in all branches of mathematics and it is used extensively in applied mathematics and Computer Graphics.

Linear Algebra is the study of vector spaces, which are mathematical structures used to design aircraft models. It is a bridge connecting mathematics with various branches of computer science. We study, how problems in almost every conceivable discipline can be solved using techniques of Linear Algebra.

The aim of this textbook on Linear Algebra is to introduce basic concepts and some applications to model computational problems.

In Chapter 1, we explain the concepts of vector spaces, subspaces , Null spaces, Row spaces and Column spaces. They are illustrated with the help of a number of useful and interesting examples.The dimension of a vector space and rank of a matrix is explained at the end of the chapter.

Chapter 2 deals with Eigenvalues, Eigenvectors,characteristic equations and diagonalization process. The connection of eigenvectors and linear transformations is also explained with the help of examples. Orthogonality of vectors is the core part of chapter 3. The concepts of inner product,length and orthogonal projections is discussed with reference to n-dimensional real vector spaces.The quadratic forms of symmetric matrices is introduces at the end of this chapter.

The fourth chapter focusses on the geometric visualization of real vector spaces with the help of affine combination of vectors and convex sets.

We are very much thankful to **Mr. Dinesh Furia** and **Mr. Jignesh Furia**, Nirali Prakashan, Pune, for valuable cooperation and guidance. Also, we are specially thankful to Mrs. Nanda Takale for her valuable cooperation. All authors dedicate this book to their parents.

We welcome any opinions and suggestions which will improve the future editions and help readers in future.

In case of queries/suggestions, send an email to: **kalyanraotakale@rediffmail.com**

Syllabus PAPER-I: MTC-121: Linear Algebra

1. Vector Spaces [10]

    1.1 Vector spaces and subspaces.

    1.2 Null spaces, column spaces and linear tranformations

    1.3 Linearly independent sets : Bases

    1.4 Coordinate systems

    1.5 The dimension of a vector space,

    1.6 Rank

2. Eigen values and Eigen vectors [10]

    2.1 Eigen values and Eigen vectors

    2.2 The characteristic equation

    2.3 Diagonalization.

    2.4 Eigen vectors and Linear transformations

3. Orthogonality and Symmetric Matrices [10]

    3.1 Inner product, length and orthogonality.

    3.2 Orthogonal sets.

    3.3 Orthogonal Projections.

    3.4 Diagonalization of Symmetric Matrices

    3.5 Quadratic forms

4. The Geometry of vector spaces [06]

    4.1 Affine combinations.

    4.2 Affine independence.

    4.3 Convex combinations.

Contents

Chapter 1

Vector Spaces

1.1 Introduction

In the last term we have learned **Matrix Theory**. It is a study of vectors in $\mathbb{R}^2, \mathbb{R}^3$ and $\mathbb{R}^n$. The vectors in $\mathbb{R}^n$ satisfy a set of properties with respect to two operations, viz. **vector addition** and **scalar multiplication**. This fact is generalised to a non-empty set of vectors **V** and the above mentioned two vector operations. The mathematical system so formed is nothing but a vector space. In reality, a space shuttle's control system works with a notion of a vector space. The set of all possible inputs of such a control system is a vector space of real-valued functions. Other examples are **Geometric Vectors** and **matrices of any fixed** order etc.

In this chapter, we study vector spaces, subspaces, null spaces, column spaces, Basis of a vector space, dimension and rank. These are elaborated and illustrated with the help of examples.

1.2 Vectors spaces and subspaces:

Definition 1.1. A vector space:

*Let **V** be a non-empty set of objects, called **vectors**. Define two operations on elements of set **V** called **vector addition** and **scalar multiplication** subject to the ten axioms given below.*

*Let $u, v, w \in$ **V** and c, d are scalars (real numbers).*

(1) The sum $\mathbf{u} + \mathbf{v} \in \mathbf{V}$. (closure under addition)

(2) $\mathbf{u} + \mathbf{v} = \mathbf{v} + \mathbf{u}$ (commutativity of addition)

(3) $\mathbf{u} + (\mathbf{v} + \mathbf{w}) = (\mathbf{u} + \mathbf{v}) + \mathbf{w}$ (Associativity of addition)

(4) A zero vector, $\mathbf{0} \in V$ such that $\mathbf{0} + \mathbf{u} = \mathbf{u} = \mathbf{u} + \mathbf{0}$ (Existence of additive identity)

(5) $\forall\, \mathbf{u} \in \mathbf{V}, \exists\, -\mathbf{u} \in \mathbf{V}$ such that $\mathbf{u} + (-\mathbf{u}) = \mathbf{0} = (-\mathbf{u}) + \mathbf{u}$

(Existence of inverse element w.r.t. addition)

(6) $\qquad$ **V** (closure under scalar multiplication)

(7) $\mathbf{c}.(\mathbf{u}+\mathbf{v}) = \mathbf{c}.\mathbf{u}+\mathbf{c}.\mathbf{v}$ (Distributivity of scalar multiplication over vector addition)

(8) $(\mathbf{c}+\mathbf{d}).\mathbf{u} = \mathbf{c}.\mathbf{u}+\mathbf{d}.\mathbf{u}$ (Distributivity of vectors over addition of Scalars)

(9) $\mathbf{c}.(\mathbf{d}.\mathbf{u}) = (\mathbf{c}.\mathbf{d}).\mathbf{u}$ (Associativity)

(10) $\mathbf{1}.\mathbf{u} = \mathbf{u}$ (Existence of an identity element for scalar multiplication)

The set **V** defined above is called a vector space whenever the ten axioms hold good for all vectors of **V**.

Now, we discuss the various examples of vector spaces.

Example 1.1. *Let $V = \mathbb{R}^2 = \left\{ \begin{bmatrix} x \\ y \end{bmatrix} \Big/ x, y \in \mathbb{R} \right\}$, V is a set of vectors in xy-plane is a real vector space.*

Solution: Let $u, v \in V$, $c, d \in \mathbb{R}$, then
$$u = \begin{bmatrix} x_1 \\ y_1 \end{bmatrix}, \quad v = \begin{bmatrix} x_2 \\ y_2 \end{bmatrix}, \quad x_1, y_1, x_2, y_2 \in \mathbb{R}.$$

Then $u + v = \begin{bmatrix} x_1 + x_2 \\ y_1 + y_2 \end{bmatrix}$ is a vector in xy-plane.

Also, $u + v = v + u$ (easy to check) and $u + (v + w) = (u + v) + w$ for all $u, v, w \in \mathbb{R}^2$.

For every vector $u \in V$, $-u \in V$ where $-u = \begin{bmatrix} -x_1 \\ -y_1 \end{bmatrix}$.

We see that $u + (-u) = \begin{bmatrix} 0 \\ 0 \end{bmatrix} \in V.$

For scalar, c, $c.u = \begin{bmatrix} cx_1 \\ cy_1 \end{bmatrix} \in V.$ The properties $(7) - (9)$ are easy to check.

$1 \cdot u = 1 \cdot \begin{bmatrix} x \\ y \end{bmatrix} = \begin{bmatrix} x \\ y \end{bmatrix} = u, \ \forall \ u \in V.$

Hence, V is a vector space. Geometrically it is visualized a shown in following figure.

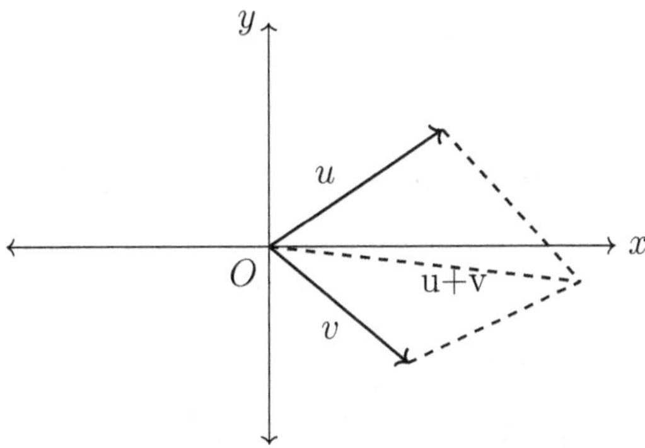

Example 1.2. *Let $V = M_{m\times n}$ be the set of all matrices of order $m \times n$ with real entries is a vector space.*

Solution: The vector addition is defined as addition of matrices and Scalar multiplication is

matrix A by c. The zero matrix is an additive identity. It is easy to see that $\forall\ A \in M_{m \times n}, (-A)$ is its additive inverse. The matrix addition is commutative and associative. The properties $(7) - (10)$ are obviously satisfied.

Example 1.3. *Let $n \geq 0$, and set $V = \mathbb{P}_n$ be the set of all polynomials of degree less equals to n is a vector space.*

Solution: Here $\mathbb{P}_n$ is the set of polynomials of degree at most n. An element $p(t)$ of $\mathbb{P}_n$ is given as follows:

$$p(t) = a_0 + a_1 t + a_2 t^2 + \cdots + a_n t^n$$

where $a_0, a_1, a_2, \cdots, a_n, t \in \mathbb{R}$.

A constant polynomial $p(t) = a_0 \neq 0$ is of degree zero. The vector $p(0) = 0$ is called a zero polynomial. This zero polynomial belongs to $\mathbb{P}_n$.

Let

$$q(t) = b_0 + b_1 t + \cdots + b_n t^n, \ \ b_0, b_1 \cdots, b_n \in \mathbb{R}$$

then $p(t) + q(t) \in \mathbb{P}_n$ since,

$$p(t) + q(t) = (a_0 + b_0) + (a_1 + b_1)t + \cdots + (a_n + b_n)t^n$$

Also,

$$c\, p(t) = c a_0 + c a_1 t + c a_2 t^2 + \cdots + c a_n t^n \in \mathbb{P}_n$$

Axioms $(2) - (3)$ and $(7) - (10)$ follow from properties of the real numbers. The zero polynomial acts as a zero element. Now, $(-1)p(t)$ is the additive inverse of $p(t) \in \mathbb{P}_n$.

Hence, $\mathbb{P}_n$ is a vector space.

Example 1.4. *Let $V = \{f/f : \mathbb{R} \to \mathbb{R}\}$ be the set of all real valued functions defined on a set $\mathbb{R}$ is a real vector space.*

Solution: Let $f, g \in V$ then vector addition is defined as $f(x) + g(x) \in \mathbb{R}$, scalar multiplication is defined as $c \in \mathbb{R}, \ cf(x) \in \mathbb{R}$.

The zero vector in V is given by the function $f(x) = 0, \forall\ x \in \mathbb{R}$. Now, $(-1)f \in \mathbb{R}$ is the additive inverse of every $f \in V$.

Also, $1.f = f \ \forall\ f \in V$.

The axioms (1) and (6) are obviously true.

The axioms $(7) - (9)$ are satisfied by properties of real numbers.

Hence, V is a real vector space.

Example 1.5. *Let V denote a space of discrete time signals. That is it is a space of all doubly infinite sequences of numbers written as follows*

then V is a vector space.

Solution: Here, $V = \{\{x_k\}/x_i \in \mathbb{R}, i = \cdots -2, -1, 0, 1, 2, \cdots\}$

Let $\{x_k\}, \{y_k\} \in V$ then define vector addition as

$$\{x_k\} + \{y_k\} = \{x_k + y_k\}$$

and Scalar multiplication is defined as, for

$$c \in \mathbb{R}, c\{x_k\} = \{cx_k\}.$$

It is easy to see that vector space axioms are verified with additive inverse of $\{x_k\}$ as $\{-x_k\}$, the zero sequences $\{0\}$ will be the additive identity of v, $1\{x_k\} = \{x_k\}$.

The remaining axioms are satisfied by using the properties of real numbers.

Hence, V is a vector space.

We prove the basic properties of scalar multiplication operation in a vector space in the following theorem.

Theorem 1.1. *Let V be a vector space. Let $u \in V, k \in \mathbb{R}$ Then,*

(a) $0\,u = 0$ (b) $k\,0 = 0$ (c) $(-1)u = -u$ (d) If $ku = 0$ then either $k = 0$ or $u = 0$

Proof: (a) By axiom (8), we have

$$0u + 0u = (0 + 0)u = 0u.$$

By axiom (5), additive inverse of $0u$ is $-0u$.

Therefore,

$$(-0u) + (0u + 0u) = (-0u) + (0u)$$

$$\text{i.e. } (-0u + 0u) + 0u = 0 \qquad\qquad \text{(by axiom (3))}$$

$$\text{i.e. } 0 + 0u = 0 \qquad\qquad\qquad \text{(by axiom (5))}$$

$$\text{i.e. } 0u = 0 \qquad\qquad\qquad\qquad \text{(by axiom (4))}$$

(b) we have

$$k(0 + 0) = k0 \qquad\qquad\qquad \text{(by axiom (4))}$$

$$\text{i.e. } k0 + k0 = k0 \qquad\qquad\qquad \text{(by axiom (7))}$$

$$\text{i.e. } k0 + k0 + (-k0) = k0 + (-k0) \qquad\qquad \text{(by axiom (5))}$$

$$\text{i.e. } k0 + (k0 + (-k0)) = 0 \qquad\qquad \text{(by axiom (3) and (4))}$$

$$\text{i.e. } k0 = 0 \qquad\qquad\qquad\qquad \text{(by axiom (4))}$$

(c) we have

$$u + (-1)u = 1u + (-1)u \qquad \text{(by axiom (10))}$$
$$= (1 + -1)u \qquad \text{(by axiom (8))}$$

i.e. $u + (-1)u = 0u$

i.e. $(-1)u = -u$

(d) Let $ku = 0$ then either $k = 0$ or $k \neq 0$.

If $k = 0$ then we are through. Else dividing by k on both sides we get

$$\frac{1}{k}(ku) = \frac{1}{k}(0) \text{ i.e. } (\frac{1}{k}k)u = 0$$

By axiom (9) i.e. $1.u = 0$ hence $u = 0$ (by axiom 10).

At this point, we discuss Euclidean n-space. We know that a real number line is a geometric object. To each point of a real line we associate a real number. Every point of xy-plane corresponds to an ordered pair of real number. Similarly every point in $\mathbb{R}^3$ corresponds to an ordered triplet of real numbers. When we try to visualize geometrically beyond $\mathbb{R}^3$ it becomes difficult. Here, we define ordered n-tuples of real numbers with addition and scalar multiplication in the following definition.

Definition 1.2. *Let $\mathbb{R}^n$ denote the set of all ordered n-tuples of real numbers. Two vectors $v = (v_1, v_2, \cdots, v_n)$ and $w = (w_1, w_2, \cdots, w_n)$ are called equal vectors if*

$$v_1 = w_1, v_2 = w_2, \cdots, v_n = w_n.$$

The sum $v + w$ is defined as componentwise addition as

$$v + w = (v_1 + w_1, v_2 + w_2, \cdots, v_n + w_n).$$

The scalar multiplication for any $k \in \mathbb{R}$ is defined as follows:

$$kv = k(v_1, v_2 \cdots v_n) = (kv_1, kv_2 \cdots kv_n).$$

These operations are called **standard operations on $\mathbb{R}^n$**.

Here we introduce matrix notation for vectors in $\mathbb{R}^n$. If $u \in \mathbb{R}^n$ then $u = \begin{bmatrix} u_1 \\ u_2 \\ \vdots \\ u_n \end{bmatrix}$.

Therefore, $u, v \in \mathbb{R}^n \Rightarrow u + v = \begin{bmatrix} u_1 \\ u_2 \\ \vdots \\ u_n \end{bmatrix} + \begin{bmatrix} v_1 \\ v_2 \\ \vdots \\ v_n \end{bmatrix} = \begin{bmatrix} u_1 + v_1 \\ u_2 + v_2 \\ \vdots \\ u_n + v_n \end{bmatrix}$

Also,

$$ku = k \begin{bmatrix} u_1 \\ u_2 \\ \vdots \end{bmatrix} = \begin{bmatrix} ku_1 \\ ku_2 \\ \vdots \end{bmatrix}$$

Example 1.6. *Suppose $V = \mathbb{R}^n$ is a vector space with addition of vectors defined componentwise and scalar multiplication defined as componentwise multiplication by a scalar. It is easy to see that $\mathbb{R}^n$ is a vector space. (verify).*

The following examples illustrate the way of **disproving** of a given set of vectors as a vector space.

Example 1.7. *Let $V = \left\{ \begin{bmatrix} x \\ y \end{bmatrix} \Big/ x \geq 0, y \geq 0 \right\}$ that is V is a set of vectors from first quadrant only is not a vector space.*

Solution: It is clear that if $u, v \in V$ then $u + v \in V$.

However, if $c = -1$ then for any $u \in V,\; cu \notin V$ as

$$(-1) \begin{bmatrix} x \\ y \end{bmatrix} = \begin{bmatrix} -x \\ -y \end{bmatrix}$$

whenever $x \geq 0, y \geq 0$. That is cu belongs to third quadrant. Hence scalar multiplication is not closed in V.

Hence, V is not a vector space.

Example 1.8. *Show that $H = \{(3s, 2 + 5s)/s \in \mathbb{R}\}$ is not a vector space.*

Solution: Note that $H \subseteq \mathbb{R}^2$. Let $u = \begin{bmatrix} 3 \\ 7 \end{bmatrix}$, $c = 2$.

Therefore, $cu = 2 \begin{bmatrix} 3 \\ 7 \end{bmatrix} = \begin{bmatrix} 6 \\ 14 \end{bmatrix}$

Now, $cu \in H$ if $\exists\, s \in \mathbb{R}$ such that $\begin{bmatrix} 3s \\ 2 + 5s \end{bmatrix} = \begin{bmatrix} 6 \\ 14 \end{bmatrix}$.

This implies that $s = 2$ and $s = \frac{12}{5}$ which is impossible because s is not unique.

Hence, $2u \notin H$. Therefore, H is not a vector space.

Example 1.9. *Let $y = 2x + 1$ be a given line in XY-plane. Then the set formed as*

$$T = \{(x, 2x + 1)/x \in \mathbb{R}\}$$

is a line not passing through origin is not a vector space.

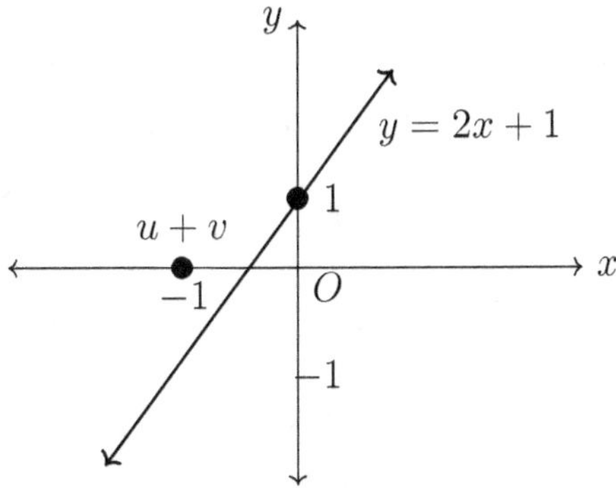

Solution: Let $u = (0, 1)$ and $v = (-1, -1) \cdot u, v \in T$. But $u + v$ does not belong to T as shown in

Example 1.10. *Suppose S is a set of all polynomials of degree 2 such that sum of their coefficients is 3. Show that S is not a vector space.*

Solution: Let $S = \{ax^2 + bx + c / a + b + c = 3, a, b, c \in \mathbb{R}\}$.

Now, S is not a vector space because for $p(x) = x^2 + 2x$ and $q(x) = 2x^2 + 1$, we observe that $p(x), q(x) \in S$, but $p(x) + q(x) = 3x^2 + 2x + 1$ does not belong to S.

Hence, the vector addition is not closed in S.

Example 1.11. *Let $S = \left\{ \begin{bmatrix} a & b \\ c & d \end{bmatrix} \middle/ b = 0, c = 1 \right\}$ and $a, d \in \mathbb{R}$ that is $S = \left\{ \begin{bmatrix} a & 0 \\ 1 & d \end{bmatrix} \middle/ a, b, d \in \mathbb{R} \right\}$ is not a vector space.*

Solution: Let $u, v \in S, u = \begin{bmatrix} 5 & 0 \\ 1 & -1 \end{bmatrix}$ and $v = \begin{bmatrix} \frac{-3}{2} & 0 \\ 1 & 1 \end{bmatrix}$.

$$\therefore u + v = \begin{bmatrix} \frac{7}{2} & 0 \\ 2 & 0 \end{bmatrix} \notin S.$$

Hence, vector addition is not closed in S. Therefore, S is not a vector space.

Example 1.12. *Let $V = \mathbb{R}^3$, define vector addition and scalar multiplication as, for $u, v \in V$ then $u + v$ is coordinate wise addition. $k \in \mathbb{R}, ku = (kx, y, z)$ where $u = (x, y, z) \in \mathbb{R}^3$.*

Solution: Then we see that vector addition is well defined and satisfies first five axioms. But the 8th axiom $(c + d)u = cu + du$ fails.

As,

$$(c + d)u = (c + d)(x, y, z) = ((c + d)x, y, z)$$

$$cu + du = (cx, y, z) + (dx, y, z) = ((c + d)x, 2y, 2z).$$

$$\therefore (c + d)u \neq cu + du.$$

Hence, V is not a vector space with given vector addition and scalar multiplication operations.

- **Subspaces**

A larger vector space may consist of an appropriate subset of vectors which is a vector space itself with respect to same operations of addition and scalar multiplication. Such a non-empty subset is known as a subspace. In this case, we require to check only three of the ten vector space axioms, the rest are automatically checked. Here, let us discuss the concept of a subspace with an example. Let

$$V = \mathbb{R}^3, \quad W = \{(x, y, 0)/x, y \in \mathbb{R}\}$$

with standard operations of addition and scalar multiplication.

Clearly, $W \subseteq V, (0, 0, 0) \in W$, for $u, v \in W$ $u + v \in W$ and for any $c \in \mathbb{R}$, $cu \in W$.

The other seven axioms are automatically satisfied by W, as they are true for a larger set V. Hence,

Definition 1.3. A subspace of a vector space.

A subset W of a vector space V that satisfies the following three properties

(a) The zero vector of V is in W i.e. $0 \in W$.

(b) W is closed under vector addition i.e. for $u, v \in W \Rightarrow u + v \in W$.

(c) W is closed under scalar multiplication i.e. for $c \in \mathbb{R}$ and $u \in V \Rightarrow cu \in W$

is known as a subspace of V.

From the above definition of a vector space we note the following valuable facts.

1. W is a non-empty subset of V as it contains a zero element of V.

2. Axioms $(2), (3)$ and $(7) - (10)$ are automatically satisfied in W as they apply to all elements of the larger set V.

3. Axiom (5) is also satisfied by W as $(-1)u \in W$ by property (c) stated in the definition of a subspace.

4. Every subspace is a vector space and every vector space is a subspace of itself.

5. By combining (b) and (c) from the definition of a subspace we state the following result.

   A subset W is a subspace of a vector space V if and only if (a) W contains the zero vector of V. and (b) for all $c, d \in \mathbb{R}$ and $u, v \in W$, $cu + dv \in W$.

Example 1.13. *Let $W = \{0\}$, where 0 is the zero vector of V. Then W is a subspace of V, it is called a zero subspace of V.*

Example 1.14. *Suppose $V = \mathbb{R}^2, W$ is a line passing through origin $(0,0)$. Then W is a subspace of $\mathbb{R}^2$. (verify)*

Example 1.15. *A plane in $\mathbb{R}^3$ passing through origin is a subspace of $\mathbb{R}^3$.*

Example 1.16. *Suppose $V = M_{m \times n}$ a set of matrices of order $m \times n$ with real entries. We define $W = \left\{ \begin{bmatrix} a & b \\ 0 & d \end{bmatrix} \middle/ a, b, d \in \mathbb{R} \right\}$, show that W is a subspace of V.*

Solution: We have $\begin{bmatrix} 0 & 0 \\ 0 & 0 \end{bmatrix} \in W$.

Let $A, B \in W$, that is $A = \begin{bmatrix} a_1 & b_1 \\ 0 & d_1 \end{bmatrix}$, $B = \begin{bmatrix} a_2 & b_2 \\ 0 & d_2 \end{bmatrix}$.

Then $A + B = \begin{bmatrix} a_1 + a_2 & b_1 + b_2 \\ 0 & d_1 + d_2 \end{bmatrix} \in W$.

Therefore, vector addition is closed in W.

Similarly,

$$p \in \mathbb{R}, \quad pA = \begin{bmatrix} pa & pb \\ 0 & pd \end{bmatrix} \in W.$$

Example 1.17. *Show that $W = \{(a, b, c)/b = a + c\}$ is a subspace of $\mathbb{R}^3$.*

Solution: We observe that the element $(1, 2, 1) \in W$. So $W \neq \phi$.

Let $u, v \in W$ and $u = (a_1, b_1, c_1), v = (a_2, b_2, c_2)$ then $b_1 = a_1 + c_1$ and $b_2 = a_2 + c_2$.

$$u + v = (a_1 + a_2, b_1 + b_2, c_1 + c_2),$$

but

$$b_1 + b_2 = (a_1 + a_2) + (c_1 + c_2).$$

Hence, $u + v \in W$.

Also, for $p \in \mathbb{R}$, $pu = (pa_1, pb_1, pc_1)$ where $pb_1 = pa_1 + pc_1 = p(a_1 + c_1)$.

This shows that $pu \in W$.

Hence, W is a subspace of V.

Example 1.18. *Let $V = \mathbb{P}_n, n \geq 0, W = \{p(t)/p(t) = at^2, a \in \mathbb{R}\}$ and W is a set of all polynomials at^2 with $a \in \mathbb{R}$. Show that W is a subspace of V.*

Solution: Note that when $a = 0, 0t^2 \in W$, so zero vector belongs to W.

Let $u, v \in W, u = a_1 t^2, v = a_2 t^2$.

We have

$$u + v = a_1 t^2 + a_2 t^2 = (a_1 + a_2)t^2 \in W.$$

Also, for $c \in \mathbb{R}, cp(t) = cu = c(a_1 t^2) = (ca_1)t^2 \in W$.

Hence, W is a subspace of V. In fact W is generated by t^2.

Example 1.19. $V = \mathbb{R}^3, W = \left\{ \begin{bmatrix} 2t \\ 0 \\ -t \end{bmatrix} \middle/ t \in \mathbb{R} \right\}$. *Show that W is a subspace of V.*

Solution: For $t = 0$ zero vector is in W.

Let

$$u, v \in W, \quad u = \begin{bmatrix} 2t_1 \\ 0 \\ -t_1 \end{bmatrix}, \quad v = \begin{bmatrix} 2t_2 \\ 0 \\ -t_2 \end{bmatrix}, \quad t_1, t_2 \in \mathbb{R}.$$

Here, $u + v = \begin{bmatrix} 2(t_1 + t_2) \\ 0 \\ -(t_1 + t_2) \end{bmatrix} \in W.$

It is easy to see that for $c \in \mathbb{R}, cu \in W$.

Hence, W is a subspace of V. In fact, the set of all linear combinations of vector

$$\begin{bmatrix} 2t \\ 0 \\ -t \end{bmatrix}$$

is nothing but space W defined above.

That is Span $\left\{ \begin{bmatrix} 2t \\ 0 \\ -t \end{bmatrix} \right\}$ is a subspace of V.

The following theorem states this fact precisely.

Theorem 1.2. *If $v_1, v_2, \cdots, v_p$ are vector in a vector space V, then Span $\{v_1, v_2, \cdots, v_p\}$ is a subspace of V.*

Example 1.20. *Let V be a vector space. $v_1, v_2 \in V$. Let $S = \{v_1, v_2\}$, then span (S) forms a subspace of V.*

Solution: We know that span (S) is the set of all vectors that can be written as a linear combination of vector of S.

The zero vector $0 = 0v_1 + 0v_2$, hence $0 \in$ Span (S).

Let $u, v \in$ Span (S) then $u = s_1v_1 + s_2v_2$ and $v = t_1v_1 + t_2v_2$.

Therefore, $u + v = (s_1 + t_1)v_1 + (s_2 + t_2)v_2 \in$ Span (S).

Let $c \in \mathbb{R}$, $cu = c(s_1v_1 + s_2v_2) = (cs_1)v_1 + (cs_2)v_2$, So, $cv \in$ Span (S). This shows that Span (S) is closed under vector addition and scalar multiplication.

Thus Span (S) is a subspace of V.

Example 1.21. *Let $T = \{(a - 3b, b - a, a, b)/a, b \in \mathbb{R}\}$.*
Show that T is a subspace of vector space $\mathbb{R}^4$ by using theorem (1.2).

Solution: Let $v \in T$ then

$$v = \begin{bmatrix} a - 3b \\ b - a \\ a \\ b \end{bmatrix} = a \begin{bmatrix} 1 \\ -1 \\ 1 \\ 0 \end{bmatrix} + b \begin{bmatrix} -3 \\ 1 \\ 0 \\ 1 \end{bmatrix}$$

Let $v_1 = \begin{bmatrix} 1 \\ -1 \\ 1 \\ 0 \end{bmatrix}$ and $v_2 = \begin{bmatrix} -3 \\ 1 \\ 0 \\ 1 \end{bmatrix}$.

The vector v is in the span $\{v_1, v_2\}$. That is the set T is nothing but span $\{v_1, v_2\}$.

Hence, by theorem (1.2), T is a subspace of $\mathbb{R}^4$.

Example 1.22. *For what value(s) of k will the vector w be in the subspace of $\mathbb{R}^3$ spanned by v_1, v_2, v_3 given below.*

$$v_1 = \begin{bmatrix} 1 \\ -1 \\ -2 \end{bmatrix}, \quad v_2 = \begin{bmatrix} 5 \\ -4 \\ -7 \end{bmatrix}, \quad v_3 = \begin{bmatrix} -3 \\ 1 \\ 0 \end{bmatrix}, \quad w = \begin{bmatrix} -4 \\ 3 \\ k \end{bmatrix}.$$

Solution: Here, the subspace spanned by vectors v_1, v_2 and v_3 is Span $\{v_1, v_2, v_3\}$. The vector $w \in$ Span $\{v_1, v_2, v_3\}$ if and only if the vector equation

$$x_1 \begin{bmatrix} 1 \\ -1 \\ -2 \end{bmatrix} + x_2 \begin{bmatrix} 5 \\ -4 \\ -7 \end{bmatrix} + x_3 \begin{bmatrix} -3 \\ 1 \\ 0 \end{bmatrix} = \begin{bmatrix} -4 \\ 3 \\ k \end{bmatrix}$$

is consistent.

That is

$$\begin{bmatrix} 1 & 5 & -3 & -4 \\ -1 & -4 & 1 & 3 \\ -2 & -7 & 0 & k \end{bmatrix} \sim \begin{bmatrix} 1 & 5 & -3 & -4 \\ 0 & 1 & 2 & -1 \\ 0 & 0 & 0 & k-5 \end{bmatrix}$$

There is no pivot element in the fourth column. Therefore $k - 5 = 0$.

Hence, $w \in$ Span $\{v_1, v_2, v_3\}$ iff $k = 5$.

Example 1.23. *The set S of all symmetric matrices of order 3×3 is a subspace of the vector sapce $M_{3\times3}$ with usual notation.*

Solution: We know that a matrix of order $n \times n$ is symmetric if $A^T = A$.

Note that a zero matrix of order 3×3 is a symmetric matrix. So, zero element (zero matrix) belongs to set S. Let $u, v \in S$ then $u^T = u$ and $v^T = v$.

Therefore, $(u + v)^T = u^T + v^T$ by property of transpose matrices. Thus $u, v \in S$. Also, it is easy to see that $c\,u \in S$, where $c \in \mathbb{R}$. Hence set S is a subspace of $M_{3\times3}$.

At this point now, we discuss some examples of sets of vectors that are not subspaces of the respective vector spaces.

Example 1.24. *The vector space $\mathbb{R}^2$ is not a subspace of $\mathbb{R}^3$. In fact it is not even a subset of $\mathbb{R}^3$.*

Example 1.25. *A plane in $\mathbb{R}^3$ that does not pass through origin is not a subspace of $\mathbb{R}^3$. This is true since the plane does not contain a zero vector. Similarly, a line in $\mathbb{R}^2$ not passing through the origin is not a subspace of $\mathbb{R}^2$.*

Example 1.26. *Let $S = \{(x, y)/y = 2\}, V = \mathbb{R}^2$. Then $(1, 2) \in S, (2, 2) \in S$ but*

$$(1, 2) + (2, 2) = (3, 4) \notin S.$$

Hence, S is not closed for vector addition, hence, S is not a subspace of V.

Example 1.27. *Show that $S = \{(x_1, x_2, \cdots, x_n)/x_n = 1\}, V = \mathbb{R}^n$ is not a subspace of V.*

Solution: Now, S is not a subspace of V because

$$u = (0, 0, 0 \cdots, 1), v = (1, 1, 1 \cdots, 1) \in S$$

but $\quad + \quad = (1 \ 1 \ 1 \quad 2) \notin S$

Example 1.28. *Is $S = \{(x, y)/xy = 1\}$, $V = \mathbb{R}^2$ a subspace of V? Justify your answer.*

Solution: No.

The zero element $O = (0, 0)$ is not in set S because $0.0 = 0 \neq 1$.

Example 1.29. *Is $S = \left\{ \begin{bmatrix} a & 0 \\ 1 & d \end{bmatrix} \middle/ a, d \in \mathbb{R} \right\}$, $V = M_{2\times 2}$ a subspace of V?*

Solution: Let $A = \begin{bmatrix} a_1 & 0 \\ 1 & d_1 \end{bmatrix}$, $B = \begin{bmatrix} a_2 & 0 \\ 1 & d_2 \end{bmatrix}$ then

$$A + B = \begin{bmatrix} a_1 + a_2 & 0 \\ 2 & d_1 + d_2 \end{bmatrix} \notin S.$$

Hence, S is not closed for vector addition. Therefore, S is not a subspace of V.

Example 1.30. *Is $S = \{(x, y)/x, y \in \mathbb{R}\} = \mathbb{R}^2$ a vector space? Justify your answer.*

Solution: No. We define vector addition as coordinate wise addition and scalar multiplication $k(x, y)$ as $(kx, 0)$. Here we check the $(10)^{\text{th}}$ axiom of a vector space $(1 \cdot u = u)$ with the given definition as follows:

$$1(x, y) = (x, 0) \neq (x, y).$$

Hence, S is not a vector space.

Example 1.31. *Prove that intersection of two subspaces is a subspace.*

Solution: Let H and K be two subspaces of a vector space V. Let $u, v \in H \cap K$ this implies that $u, v \in H$ and $u, v \in K$. But H and K are themself subspaces. So $u + v \in H$ and $u + v \in K$ Hence $u + v \in H \cap K$. Also, for vector $u \in H \cap K, cu \in H$ and $cu \in K$ for some $c \in \mathbb{R}$. Therefore scalar multiplication is clossd in $H \cap K$. Hence $H \cap k$ is a subspace of V.

Example 1.32. *Show that union of two subspaces may not be a subspace of a vector space.*

Solution: Let

$$S_1 = \left\{ \begin{bmatrix} a & 0 \\ 0 & d \end{bmatrix} \middle/ a, d \in \mathbb{R} \right\}, S_2 = \left\{ \begin{bmatrix} a & a - d \\ c & d \end{bmatrix} \middle/ a, c, d \in \mathbb{R} \right\}.$$

S_1 and S_2 are subspaces of $M_{2\times 2}$ with usual operation of vector addition and scalar multiplication on matrices. Let

$$\begin{bmatrix} 9 & 0 \\ 0 & 1 \end{bmatrix} \in S_1 \text{ and } \begin{bmatrix} 1 & -3 \\ 3 & 4 \end{bmatrix} \in S_2$$

$$\begin{bmatrix} 9 & 0 \\ 0 & 1 \end{bmatrix} + \begin{bmatrix} 1 & -3 \\ 3 & 4 \end{bmatrix} = \begin{bmatrix} 10 & -3 \\ 0 & 5 \end{bmatrix} \notin S_1 \cup S_2.$$

Therefore, vector addition is not closed in $S_1 \cup S_2$.

1.3 Null spaces, Column spaces and Linear transformation

This section, is a study of special and important subspaces of $\mathbb{R}^n$ because subspaces play an important role in applications of linear algebra. The subspaces of $\mathbb{R}^n$ are of two types, one as a set of all solution to a system of homogeneous linear equations and secondly, as a set of linear combinations of certain specified vectors.

Definition 1.4. *The **Null space** of an $m \times n$ matrix A, written as Nul A, is the set of all solutions of the homogeneous equation $Ax = O$.*

That is, Nul $A = \{x/x \in \mathbb{R}^n, Ax = O\}$.

Actually the Nul A can be described as set of all $x \in \mathbb{R}^n$ that are mapped onto a zero vector of $\mathbb{R}^m$ via a linear transformation $x \mapsto Ax$ as shown in the figure.

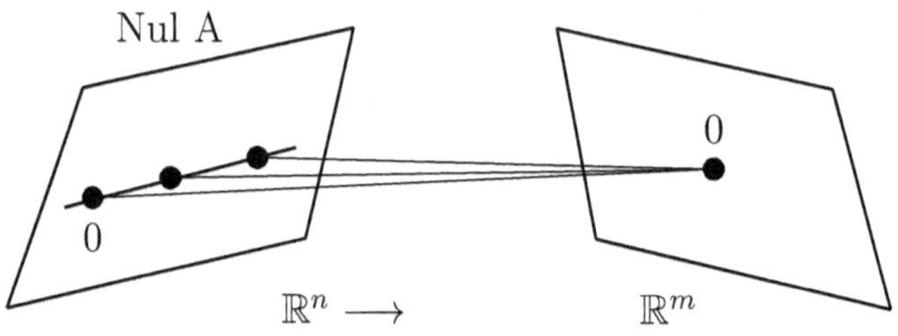

Theorem 1.3. *The Nul A is a subspace of $\mathbb{R}^n$.*

Proof: It is clear that Nul $A \subseteq \mathbb{R}^n$. The zero vector belongs to Nul A by its definition.

Let $u, v \in$ Nul A. $\therefore Au = 0$ and $Av = 0$

Consider $A(u + v) = Au + Av = O + O = O$ (By property of matrix multiplication)

Let, $c \in \mathbb{R}$, then $A(cu) = c(Au) = cO = O$.

This shows that $u + v \in$ Nul A and $cu \in$ Nul A.

Hence, Nul A is a subspace of $\mathbb{R}^n$.

Remark 1.1. *We have already learnt the method of solving a homogeneous system. Now, we will explicitly describe Nul A for matrix A. Also, we determine whether the given vector belongs to Nul A or not.*

Example 1.33. *Let*

$$w = \begin{bmatrix} 1 \\ 3 \\ -4 \end{bmatrix}, \quad A = \begin{bmatrix} 3 & -5 & -3 \\ 6 & -2 & 0 \\ -8 & 4 & 1 \end{bmatrix}$$

For given vector w we determine whether w is in the Nul A. Comput Aw as follows:

$$\begin{bmatrix} 3 & -5 & -3 \\ 6 & -2 & 0 \\ -8 & 4 & 1 \end{bmatrix} \begin{bmatrix} 1 \\ 3 \\ -4 \end{bmatrix} = \begin{bmatrix} 0 \\ 0 \\ 0 \end{bmatrix}$$

Example 1.34. *Let* $A = \begin{bmatrix} 1 & -6 & 4 & 0 \\ 0 & 0 & 2 & 0 \end{bmatrix}.$
For matrix A we find Nul A.

Reduce matrix A to row echelon form.

$$\begin{bmatrix} 1 & -6 & 4 & 0 \\ 0 & 0 & 2 & 0 \end{bmatrix} \sim \begin{bmatrix} 1 & -6 & 4 \\ 0 & 0 & 2 \end{bmatrix}$$

The given matrix A is in row echelon form with two leading elements, as x and z (say).

$$\text{So, } x = 6y - 4z, 2z = 0$$

$$\therefore z = 0 \Rightarrow x = 6y$$

$\therefore$ The general solution is

$$\begin{bmatrix} x \\ y \\ z \end{bmatrix} = \begin{bmatrix} 6y \\ y \\ 0 \end{bmatrix} = y \begin{bmatrix} 6 \\ 1 \\ 0 \end{bmatrix} = yu$$

This shows that every linear combination of vector u is an element of Nul A. That is, the set $\{u\}$ Spans Nul A. Hence Nul $A = \text{Span } \{u\}$.

Example 1.35. *Find NulA where* $A = \begin{bmatrix} 1 & 5 & -4 & -3 & 1 \\ 0 & 1 & -2 & 1 & 0 \\ 0 & 0 & 0 & 0 & 0 \end{bmatrix}.$

Solution: Observe that matrix A is in row echelon form. The leading elements are x and y (say). There are three free variables z, w and t (say). The given homogeneous system can be written as

$$x + 5y - 4z - 3w + t = 0$$

$$y - 2z + w = 0$$

$$\therefore x = -5y + 4z + 3w - t$$

$$y = 2z - w$$

Hence, the general solution is

$$\begin{bmatrix} x \\ y \\ z \\ w \\ t \end{bmatrix} = \begin{bmatrix} -6z + 8w - t \\ 2z - w \\ z \\ w \\ t \end{bmatrix} = z \underset{\underset{v_1}{\uparrow}}{\begin{bmatrix} -6 \\ 2 \\ 1 \\ 0 \\ 0 \end{bmatrix}} + w \underset{\underset{v_2}{\uparrow}}{\begin{bmatrix} 8 \\ -1 \\ 0 \\ 1 \\ 0 \end{bmatrix}} + t \underset{\underset{v_3}{\uparrow}}{\begin{bmatrix} -1 \\ 0 \\ 0 \\ 0 \\ 1 \end{bmatrix}}$$

$$= zv_1 + wv_2 + tv_3$$

This shows that every linear combination of vectors v_1, v_2, v_3 is an element of Nul A. That is the set

Example 1.36. *Let* $A = \begin{bmatrix} 2 & -6 \\ -1 & 3 \\ -4 & 12 \\ 3 & -9 \end{bmatrix}$. *Find k such that Nul A is a subspace of $\mathbb{R}^k$.*

Solution: Note that a vector x such that Ax is defined must have 2 entries. So Nul A is a subspace of $\mathbb{R}^k$, where $k = 2$.

We discuss the column space of a matrix A. It is defined explicitly via linear combination of columns of A.

Definition 1.5. *The* **column space** *of an $m \times n$ matrix A, written as Col A, is the set of all linear combination of the columns of A. If $A = [a_1 a_2 \cdots a_n]$ then Col $A = Span\ \{a_1, a_2, \cdots, a_n\}$. We know that $Span\ \{a_1, a_2, \cdots, a_n\}$ is a subspace by theorem (1.2).*

Theorem 1.4. *The column space of an $m \times n$ matrix A is a subspace $\mathbb{R}^m$*

Proof: We know that Col A ia a subspace by theorem (1.1). Also any vector in Col A can be written as Ax for some $x \in \mathbb{R}^n$. Ax is a linear combination of column of matrix A. That is

$$\text{Col } A = \{b/b = Ax, x \in \mathbb{R}^n\}.$$

The vector b in Col A are of the from $\begin{bmatrix} b_1 \\ b_2 \\ \vdots \\ b_m \end{bmatrix} \in \mathbb{R}^m$.

Hence, Col A is a subspace of $\mathbb{R}^m$.

Example 1.37. *Find matrix A such that* $W = \left\{ \begin{bmatrix} b - c \\ 2b + c + d \\ 5c - 4d \\ d \end{bmatrix} \middle/ b, c, d \in \mathbb{R} \right\} = Col A$

Solution: We find matrix A such that $W = \text{Col } A$

$$\begin{bmatrix} b - c \\ 2b + c + d \\ 5c - 4d \\ d \end{bmatrix} = b \begin{bmatrix} 1 \\ 2 \\ 0 \\ 0 \end{bmatrix} + c \begin{bmatrix} -1 \\ 1 \\ 5 \\ 0 \end{bmatrix} + d \begin{bmatrix} 0 \\ 1 \\ -4 \\ 1 \end{bmatrix}$$

Therefore,

$$W = \left\{ b \begin{bmatrix} 1 \\ 2 \\ 0 \\ 0 \end{bmatrix} + c \begin{bmatrix} -1 \\ 1 \\ 5 \\ 0 \end{bmatrix} + d \begin{bmatrix} 0 \\ 1 \\ -4 \\ 1 \end{bmatrix} \middle/ b, c, d \in \mathbb{R} \right\}$$

$$\quad\uparrow \qquad\qquad \uparrow \qquad\qquad \uparrow$$
$$\quad a_1 \qquad\qquad a_2 \qquad\qquad a_3$$

$$W = Span\ \{a_1, a_2, a_3\}$$

The vector in set W are nothing but the columns of matrix A such that $W = $ Col A.

Therefore, the required matrix $A = \begin{bmatrix} 1 & -1 & 0 \\ 2 & 1 & 1 \\ 0 & 5 & -4 \end{bmatrix}$

Example 1.38. *Let*

$$A = \begin{bmatrix} -8 & -2 & -9 \\ 6 & 4 & 8 \\ 4 & 0 & 4 \end{bmatrix}, w = \begin{bmatrix} 2 \\ 1 \\ -2 \end{bmatrix}$$

For given A and w, determine whether w is in Col A, in Nul A?

Solution: We compute

$$Aw = \begin{bmatrix} -8 & -2 & -9 \\ 6 & 4 & 8 \\ 4 & 0 & 4 \end{bmatrix} \begin{bmatrix} 2 \\ 1 \\ -2 \end{bmatrix} = \begin{bmatrix} 0 \\ 0 \\ 0 \end{bmatrix}$$

This shows that w is a solution of the matrix equation $Ax = O$.

Hence, $w \in$ Nul A. Now, to check whether $w \in$ Col A we solve the system $Ax = W$.

The augmented matrix is

$$\begin{bmatrix} -8 & -2 & -9 & 2 \\ 6 & 4 & 8 & 1 \\ 4 & 0 & 4 & -2 \end{bmatrix}$$

It reduces to row echelon form as

$$\begin{bmatrix} 4 & 0 & 4 & -2 \\ 0 & 4 & 2 & 4 \\ 0 & 0 & 0 & 0 \end{bmatrix}.$$

From the above form of augmented matrix it is clear that the system $Ax = w$ is consistent. Hence, vector w can be written as a linear combination of columns of A. This implies that $w \in$ Col A.

We have discussed Nul A and Col A of an $m \times n$ matrix A. At this point we compare them. The following table describes the contrast between Nul A and Col A.

Nul A	Col A
It is a set of all solutions of the matrix equation $Ax = O$	It is a set of all linear combinations of column of A.
It is subspace of $\mathbb{R}^n$	It is a subspace of $\mathbb{R}^m$
Let $v \in$ Nul A then $Av = O$.	Let $v \in$ Col A then the matrix equation $Ax = v$ is consistent.
Nul $A = \{0\}$ iff the linear transformation $x \mapsto Ax$ is one-to-one.	Col $A = \mathbb{R}^m$ if the linear transformation $x \mapsto Ax$ maps $\mathbb{R}^n$ onto $\mathbb{R}^m$.

In the last term, in Matrix Algebra course we have learned about a linear transformation $x \longmapsto Ax$ with reference to $\mathbb{R}^n \to \mathbb{R}^m$ mapping. It is a transformation of vector x to a vector Ax. Now, here, we define a linear transformation between two vector spaces V and W as follows.

Definition 1.6. *A **linear Transformation***

$T : V \to W$ (V and W are vector spaces) is a rule that assigns to each vector x in V a unique vector $T(x)$ in W such that

(a) $T(u + v) = T(u) + T(v), u, v \in V$

(b) $T(cu) = cT(u), \forall u \in V$ and $c \in \mathbb{R}$.

We know that every matrix transformation is a linear transformation.

For a matrix transformation $T : V \to W$ defined as $x \longmapsto Ax$ the Null space of T is the set of all those vectors of V that map onto O under T. That is $\{x/T(x) = 0, x \in V\}$. This set is known as **Kernel of** T.

Definition 1.7. Kernel of a linear transformation T

Let T be a linear transformation from a vector space V to a vector space W, then it is a set of all vectors $u \in V$ such that $T(u) = 0$ (the zero vector of W)

$$\therefore \ker T = \{u \in V/T(u) = 0\}.$$

Definition 1.8. The Range of a linear transformation T

Let T be a linear transformation from a vector space V to W, then it is the set of all vectors in W of the form $T(x)$ for some $x \in V$

$$\therefore \ Range(T) = \{T(x)/x \in V\}.$$

In particular if T is a matrix transformation then range of T is Col A.

Example 1.39. *If T is a matrix transformation from V to W then kernel (T) is a subspace of V.*

Solution: $\ker(T) = \{u \in V/T(u) = 0\}$

Let $u, v \in \ker(T)$ and $c, d \in \mathbb{R}$, then

$$T(cu + dv) = cT(u) + dT(v) = 0.$$

This implies that the vector $cu + dv \in \ker(T)$.

Hence, $\ker(T)$ is a subspace of V.

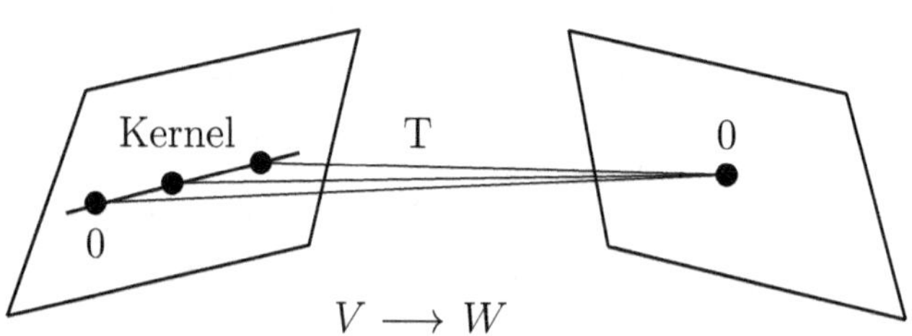

Note that, Range (T) is a subspace of W, this is true since $T(0) = 0$ and $T(cu + dv)$ is in the image set of T whenever $u, v \in$ Range (T).

Example 1.40. *Let $T : \mathbb{P}_2 \to \mathbb{R}^2$ as $T(P) = \begin{bmatrix} P(0) \\ P(1) \end{bmatrix}$, show that map T is a linear transformation.*

Solution: Let $P_1, P_2 \in \mathbb{P}_2, c, d \in \mathbb{R}$

$$\therefore T(P_1) = \begin{bmatrix} P_1(0) \\ P_1(1) \end{bmatrix} \text{ and } T(P_2) = \begin{bmatrix} P_2(0) \\ P_2(1) \end{bmatrix}$$

$$\therefore T(cP_1 + dP_2) = \begin{bmatrix} cP_1(0) + dP_2(0) \\ cP_1(1) + dP_2(1) \end{bmatrix} = c\begin{bmatrix} P_1(0) \\ P_1(1) \end{bmatrix} + d\begin{bmatrix} P_2(0) \\ P_2(1) \end{bmatrix} = cT(P_1) + dT(P_2)$$

Hence, T is a linear transformation.

$$\ker(T) = \left\{ P \in \mathbb{P}_2 \Big/ T(P) = \begin{bmatrix} 0 \\ 0 \end{bmatrix} \right\} = \left\{ P \in \mathbb{P}_2 \Big/ \begin{bmatrix} P(0) \\ P(1) \end{bmatrix} = \begin{bmatrix} 0 \\ 0 \end{bmatrix} \right\}$$

$$= \text{set of all those polynomials of } \mathbb{P}_2 \text{ whose roots are 0 and 1.}$$

A degree 2 polynomial with roots 0 and 1 is $t(t-1) = t^2 - t$.

Hence, Span $\{t^2 - t\} = \ker(T)$

Range of $T = \mathbb{R}^2$ (easy to verify)

1.4 Linearly independent sets and Bases

In this section, we identify , search and justify subsets of a vector space (or a subspace) that "**span**" the spaces and have the property of linear independence. A linearly independent set of vectors of a vector space is defined as follows.

Definition 1.9. *An indexed set of vectors* $\{v_1, v_2, \cdots v_p\}$ *in a vector space* V *is said to be* **linearly independent** *if the vector equation*

$$c_1 v_1 + c_2 v_2 + \cdots + c_p v_p = 0$$

has only a trivial solution $c_1 = c_2 = \cdots = c_p = 0.$

The set of vectors that is **not** linearly independent is **linearly dependent**.

That is the vector equation $c_1 v_1 + c_2 v_2 + \cdots + c_p v_p = 0$ has a **non-trivial** solution. This means that the coefficients (constant) $c_1, c_2, \cdots c_p$ are **not all zero** such that the corresponding equation is satisfied. The following theorem is a characterization of linearly dependent set.

Theorem 1.5. *An indexed set* $v_1, v_2, \cdots v_p$ *of two or more vectors, with* $v_1 \neq 0$ *is linearly dependent if and only if some* v_j *(with* $j > 1$*) is a linear combination of the preceding vectors* $v_1, v_2 \cdots v_{j-1}$

As a consequence of the above theorem, in case of a vector space $\mathbb{R}^n$, the following statements are useful to solve the problems.

- **A set containing single vector** $v \in \mathbb{R}^n$ **is linearly independent iff** $v \neq 0$.
- **A set of two vectors in** $\mathbb{R}^n$ **is linearly dependent iff one of the vectors is a multiple of the other.**

Example 1.41. *Let $V = \mathbb{P}_2$, $S = \{p_1, p_2, p_3\}$ where $p_1 = 1 - t$, $p_2 = 5 + 3t - t^2$, $p_3 = 1 + 3t - t^2$, show that the set S is linearly dependent.*

Solution: The set S will be linearly dependent if the vector equation $c_1 p_1 + c_2 p_2 + c_3 p_3 = 0$ has a non-trivial solution. It is determined as follows:

$$c_1 p_1 + c_2 p_2 + c_3 p_3 = 0$$

$$\therefore c_1(1 - t) + c_2(5 + 3t - t^2) + c_3(1 + 3t - t^2) = 0$$

$$\therefore (-c_2 - c_3)t^2 + (-c_1 + 3c_2 + 3c_3)t + (c_1 + 5c_2 + c_3) = 0$$

$\therefore$ comparing coefficients on both side we get

$$\left.\begin{array}{rcl} -c_2 - c_3 &=& 0 \\ -c_1 + 3c_2 + 3c_3 &=& 0 \\ c_1 + 5c_2 + c_3 &=& 0 \end{array}\right\} \tag{1.1}$$

Solving system (1.1), we see that (1.1) has non-trivial solution.

Hence, set S is linearly dependent.

Example 1.42. *Show that the subset $S = \{(3, 2, 3), (1, 1, 1), (1, 0, 1)\}$ of $\mathbb{R}^3$ linearly dependent .*

Solution: If the vector equation

$$c_1(3, 2, 3) + c_2(1, 1, 1) + c_3(1, 0, 1) = 0$$

has only the trivial solution then set S will be linearly independent.

That is, we write the following homogeneous system

$$\left.\begin{array}{rcl} 3c_1 + c_2 + c_3 &=& 0 \\ 2c_1 + c_2 &=& 0 \\ 3c_1 + c_2 + c_3 &=& 0 \end{array}\right\} \tag{1.2}$$

The determinant of coefficient $|A|$ of (1.2) shows that, it is zero.

Hence, (1.2) has a non-trivial solution.

Therefore, the given set S is linearly dependent.

Example 1.43. *Show that the set of vector $T = \left\{ \begin{bmatrix} -2 \\ 0 \\ 1 \end{bmatrix}, \begin{bmatrix} -1 \\ 0 \\ 1 \end{bmatrix} \right\}$ in $\mathbb{R}^3$ is linearly independent.*

Solution: The set T is linearly independent as the vector equation,

$$c_1 v_1 + c_2 v_2 = 0$$

where

$$v_1 = \begin{bmatrix} -2 \\ 1 \\ 0 \end{bmatrix}, \ v_2 = \begin{bmatrix} -1 \\ 0 \\ 1 \end{bmatrix}$$

Example 1.44. *Show that the set Q containing vectors*

$$v_1 = \begin{bmatrix} 4 \\ -3 \\ 7 \end{bmatrix}, \ v_2 = \begin{bmatrix} 1 \\ 9 \\ -2 \end{bmatrix}, \ v_3 = \begin{bmatrix} 7 \\ 11 \\ 6 \end{bmatrix}$$

is linearly dependent in $\mathbb{R}^3$.

Solution: The required linear combination is

$$4v_1 + 5v_2 - 3v_3 = 0.$$

At this point, now we combine the two important concept of spanning set and linearly independent set to define a basis of a vector space as follows:

Definition 1.10. *Basis*

Let H be a subspace of a vector space V. An indexed set of vector $B = \{b_1, b_2 \cdots b_p\}$ in V is a basic for H if (i) B is a linearly independent set and

(ii) the subspace spanned by B coincides with H, that is $H = Span\{b_1, b_2, \cdots b_p\}$.

Example 1.45. *Let $A = [a_1\, a_2 \cdots a_n]$ be an $n \times n$ **invertible** matrix. Then columns of A form a basis for $\mathbb{R}^n$.*

Solution: The column vectors $a_1, a_2 \cdots a_n$ are linearly independent and $\mathbb{R}^n = $ Span $\{a_1, a_2 \cdots a_n\}$.

Example 1.46. *The column vectors of an identity matrix*

$$A = \begin{bmatrix} 1 & 0 & 0 \\ 0 & 1 & 0 \\ 0 & 0 & 1 \end{bmatrix} = [e_1\, e_2\, e_3]$$

form a basic of $\mathbb{R}^3$.

Solution: We know that A is an invertible matrix (of order 3×3). Invertibility can be checked by observing number of pivot elements (here 3 pivot elements) or by finding det A (here a non-zero determinant)

Also, any vector $v \in \mathbb{R}^3$ can be written as $v = c_1 e_1 + c_2 e_2 + c_3 e_3$, $c_1, c_2, c_3 \in \mathbb{R}$.

Hence, $\mathbb{R}^3 = $ Span $\{e_1, e_2, e_3\}$.

Therefore, the set $\{e_1, e_2, e_3\}$ forms a basis of $\mathbb{R}^3$. It is called a **standard basis** of $\mathbb{R}^3$.

Generalizing the above fact, we see that the columns of the matrix

$$I_n = \begin{bmatrix} 1 & 0 & \cdots & 0 \\ 0 & 1 & \cdots & 0 \\ 0 & 0 & \cdots & 0 \\ \vdots & \vdots & & \vdots \\ 0 & 0 & \cdots & 1 \end{bmatrix} = [e_1\, e_2 \cdots e_n]$$

Example 1.47. *Let* $V = \mathbb{R}^3$, $S = \{(2, -1, 4), (3, 6, 2), (2, 10, -4)\}$

$$= \left\{ \begin{bmatrix} 2 \\ -1 \\ 4 \end{bmatrix}, \begin{bmatrix} 3 \\ 6 \\ 2 \end{bmatrix}, \begin{bmatrix} 2 \\ 10 \\ -4 \end{bmatrix} \right\} = \{a_1, a_2, a_3\}$$

is a basis of $\mathbb{R}^3$.

Solution: Let $A = [a_1, a_2, a_3]$. We see that $\det(A) \neq 0$. So the homogeneous system $Ax = 0$ has only trivial solution. This implies that set S is linearly independent.

Now we show that the set S spans $\mathbb{R}^3$. Let $b = (b_1, b_2, b_3) \in \mathbb{R}^3$.

Consider the system $Ax = b$. Because $\det(A) \neq O$ this system is consistent for any choices of b_1, b_2, b_3. Hence given any vector $b \in \mathbb{R}^3$, it is expressible as a linear combination of vectors of set S. Hence the set S spans $\mathbb{R}^3$

Therefore, S forms a basis of $\mathbb{R}^3$. Here we observe that $\mathbb{R}^3$ can have more than one bases, but the standard basis is unique.

Now, we state a useful theorem that helps us to determine whether a subset of $\mathbb{R}^3$ with more than 'n' vectors is linearly dependent or independent.

Theorem 1.6. *If* $S = \{v_1, v_2, \cdots v_n\}$ *is a basis of a vector space* V, *then every set with* **more than** *n* **vectors** *is* **linearly dependent**.

Example 1.48. *The set* $S = \left\{ \begin{bmatrix} 1 \\ -3 \\ 0 \end{bmatrix}, \begin{bmatrix} -2 \\ 9 \\ 0 \end{bmatrix}, \begin{bmatrix} 0 \\ 0 \\ 0 \end{bmatrix}, \begin{bmatrix} 0 \\ -3 \\ 5 \end{bmatrix} \right\}$ *is linearly dependent using theorem* (1.6) *with* $V = \mathbb{R}^3$.

Example 1.49. *The set* $S = \left\{ \begin{bmatrix} -2 \\ 3 \\ 0 \end{bmatrix}, \begin{bmatrix} 6 \\ -1 \\ 5 \end{bmatrix} \right\} = \{v_1, v_2\}$ *is not a basis of* $\mathbb{R}^3$.

Solution: The given set S is linearly independent as they are **not** multiples of each other. But the set **does not** span $\mathbb{R}^3$ since there are only two pivot elements in the echelon form. (Actually we need 3 pivot elements in $\mathbb{R}^3$). Hence, the set S is **not** a basis of $\mathbb{R}^3$.

Example 1.50. *The set* $S = \left\{ \begin{bmatrix} 1 \\ 0 \\ 1 \end{bmatrix}, \begin{bmatrix} 0 \\ 0 \\ 0 \end{bmatrix}, \begin{bmatrix} 0 \\ 1 \\ 0 \end{bmatrix} \right\}$ *is not basis of* $\mathbb{R}^3$.

Solution: We know that a set containing a zero vector in a vector space is always linearly dependent. Hence S does not form a basis of $\mathbb{R}^3$

Example 1.51. *The Set* $S = \{1, t, t^2, \cdots t^n\}$ *form a standard basis of* $\mathbb{P}_n$

Solution: Here we see that every polynomial of degree less than or equal to n with real coefficients can be expressed as a linear combinations of vectors of set S. So, set S spans $\mathbb{P}_n$.

To justify that S is linearly independent. Suppose that $c_0, c_1, \cdots c_n$ satisfy,

Then equating coefficient from both sides, where $O(t)$ is a zero polynomial with all coefficients zero, we see that $c_0 = c_1 \cdots = c_n = 0$ Hence the set S is linearly independent. Therefore the set S forms a basis of $\mathbb{P}_n$.

In the following discussion we tackle with answer to the question,

"How to construct a basis from a spanning set of a vector space?"

Example 1.52. *Let* $v_1 = \begin{bmatrix} 0 \\ 2 \\ -1 \end{bmatrix}$, $v_2 = \begin{bmatrix} 2 \\ 2 \\ 0 \end{bmatrix}$, $v_3 = \begin{bmatrix} 6 \\ 16 \\ -5 \end{bmatrix}$, $H = \text{Span } \{v_1, v_2, v_3\}$.

Solution: By inspection we see that $v_3 = 5v_1 + 3v_2$.

That is the set $\{v_1, v_2, v_3\}$ is linearly dependent.

We show that span $\{v_1, v_2, v_3\} = $ span $\{v_1, v_2\}$.

It is easy to note that Span $\{v_1, v_2\} \subseteq$ Span $\{v_1, v_2, v_3\}$ for, $x \in$ span $\{v_1, v_2\}$ This implies that $x = c_1 v_1 + c_2 v_2$.

That is, $x = c_1 v_1 + c_2 v_2 + c_3 v_3$ for every x, where $c_3 = 0$.

Now, let $y \in$ span $\{v_1, v_2, v_3\}$.

$$\therefore y = d_1 v_1 + d_2 v_2 + d_3 v_3.$$

But $v_3 = 5v_1 + 3v_2$, substituting for v_3 in above equation, we get

$$y = (d_1 + 5d_3)v_1 + (d_2 + 3d_3)v_2$$

$$\therefore y \in \text{span } \{v_1, v_2\}.$$

Hence, Span $\{v_1, v_2, v_3\} \subseteq$ span $\{v_1, v_2\}$

$\therefore$ Span $\{v_1, v_2, v_3\} = $ Span $\{v_1, v_2\}$.

Therefore, the subspace H of $\mathbb{R}^3$ is spanned by set $\{v_1, v_2\}$ and the set is linearly independent.

$\therefore \{v_1, v_2\}$ forms a basis of H.

In example 1.52 we have illustrated a procedure to find a basis of the subspace spanned by the given set of vectors.

Example 1.53. *Let* $v_1 = \begin{bmatrix} 1 \\ -3 \\ 4 \end{bmatrix}$, $v_2 = \begin{bmatrix} 6 \\ 2 \\ -1 \end{bmatrix}$, $v_3 = \begin{bmatrix} 2 \\ -2 \\ 3 \end{bmatrix}$, $v_4 = \begin{bmatrix} -4 \\ -8 \\ 9 \end{bmatrix}$, $W = Span\{v_1, v_2, v_3, v_4\}$.

Show that W is a ColA formed by vectors v_1, v_2, v_3, v_4.

Solution: Using the given information we find a basis for W.

Let

$$A = \begin{bmatrix} 1 & 6 & 2 & -4 \\ -3 & 2 & -2 & -8 \\ 4 & -1 & 3 & 9 \end{bmatrix}.$$

Row reduce matrix A to obtain

$$A = \begin{bmatrix} 1 & 6 & 2 & -4 \\ 0 & 5 & 1 & -5 \\ 0 & 0 & 0 & 0 \end{bmatrix}.$$

There are two pivot elements in first two columns of A.

Hence $W = \text{Col } A$. The set $\{v_1, v_2\}$ forms a basis of W.

Note that it is not needed to find $RREF$ of A to locate the pivot columns.

Theorem 1.7. *Spanning Set Theorem*

Let $S = \{v_1, v_2 \cdots v_p\}$ be a set of vectors in a vector space V, and let $H = \text{Span } \{v_1, v_2, \cdots v_p\}$.

(a) Span $\{v_1, v_2, \cdots v_k, \cdots v_p\} = \text{Span } \{v_1, v_2, \cdots v_{k-1}, v_{k+1} \cdots v_p\}$, where v_k is a vector which is written an a linear combination of remaining vectors.

(b) If $H \neq \{0\}$, then some subset of S is a basis of H.

The following example illustrates the method to find a basis for the null space of the given matrix A.

Example 1.54. *Find the basis of $NulA$ where $A = \begin{bmatrix} 1 & 0 & -3 & 2 \\ 0 & 1 & -5 & 4 \\ 3 & -2 & 1 & -2 \end{bmatrix}.$*

Solution: Row reducing matrix A we get,

$$\begin{bmatrix} 1 & 0 & -3 & 2 \\ 0 & 1 & -5 & 4 \\ 0 & 0 & 0 & 0 \end{bmatrix}.$$

The corresponding augmented matrix of the homogeneous system $Ax = O$ is

$$\begin{bmatrix} 1 & 0 & -3 & 2 & 0 \\ 0 & 1 & -5 & 4 & 0 \\ 0 & 0 & 0 & 0 & 0 \end{bmatrix}$$

Rewriting the system in terms of free variables(x_3, x_4), we get

$$x_1 = 3x_3 - 2x_4$$

$$x_2 = 5x_3 - 4x_4$$

x_3, x_4 are free variables.

$$\therefore \begin{bmatrix} x_1 \\ x_2 \\ x_3 \\ x_4 \end{bmatrix} = \begin{bmatrix} 3x_3 - 2x_4 \\ 5x_3 - 4x_4 \\ x_3 \\ x_4 \end{bmatrix} = x_3 \underset{\underset{v_1}{\uparrow}}{\begin{bmatrix} 3 \\ 5 \\ 1 \\ 0 \end{bmatrix}} + x_4 \underset{\underset{v_2}{\uparrow}}{\begin{bmatrix} -2 \\ -4 \\ 0 \\ 1 \end{bmatrix}}.$$

The set $\{v_1, v_2\}$ spans Nul A and it is linearly independent set, So the set $\{v_1, v_2\}$ forms a basis of

Example 1.55. *Find a basis of the subspace spanned by the given set*
$S = \{(1, -3, 2), (2, 4, 1), (3, 1, 3), (1, 1, 1)\} = \{v_1, v_2, v_3, v_4\}$ *of vectors.*

Solution: Here we obtain all the dependency relations of vectors using the vector equation

$$x_1 v_1 + x_2 v_2 + x_3 v_3 + x_4 v_4 = 0$$

This homogeneous linear system is written in terms of augmented matrix as:

$$\begin{bmatrix} 1 & 2 & 3 & 1 & 0 \\ -3 & 4 & 1 & 1 & 0 \\ 2 & 1 & 3 & 1 & 0 \end{bmatrix} \sim \begin{bmatrix} 1 & 2 & 3 & 1 & 0 \\ 0 & 10 & 10 & 4 & 0 \\ 0 & 0 & 0 & \frac{1}{5} & 0 \end{bmatrix}$$

Rewriting the echelon form in linear equations we get

$$x_1 + 2x_2 + 3x_3 + x_4 = 0$$

$$10x_2 + 10x_3 + 4x_4 = 0$$

$$\frac{1}{5}x_4 = 0$$

Note that x_3 is a free variable. So, The solution of the above system is

$$x_1 = -x_3, \quad x_2 = -x_3, \quad x_3, \quad x_4 = 0$$

Hence the vector equation becomes

$$-x_3 v_1 - x_3 v_2 + x_3 v_3 + 0 v_4 = 0$$

$$\therefore x_3(-v_1 - v_2 + v_3) = 0$$

x_3 is arbitrary, put $x_3 = 1$. We get $-v_1 - v_2 + v_3 = 0$

$\therefore v_3 = v_1 + v_2$ is the required dependency equation.

This shows that v_3 is a linear combination of vectors v_1 and v_2. Hence the required basis of the subspace Spanned by $\{v_1, v_2, v_3, v_4\}$ is $\{v_1, v_2, v_4\}$.

In the next example we illustrate the method of finding the basis of Col A for given matrix A. The following theorem is used in this method.

Theorem 1.8. *The pivot columns of a matrix A form a basis for Col A.*

Example 1.56. *Find the basis of Col A where*

$$A = \begin{bmatrix} 1 & 4 & 0 & 2 & -1 \\ 3 & 12 & 1 & 5 & 5 \\ 2 & 8 & 1 & 3 & 2 \\ 5 & 20 & 2 & 8 & 8 \end{bmatrix} = \begin{bmatrix} a_1 & a_2 & a_3 & a_4 & a_5 \end{bmatrix}.$$

Solution: It is easy to see that matrix A reduces to

$$\begin{bmatrix} 1 & 4 & 0 & 2 & 0 \\ 0 & 0 & 1 & -1 & 0 \\ 0 & 0 & 0 & 0 & 1 \end{bmatrix}$$

Hence by theorem 1.7 the vectors corresponding to pivot columns are $a_1\, a_3, a_5$. Therefore the set $\{a_1, a_3, a_5\}$ forms a basis of Col A.

1.5   Coordinate Systems

In the last section we learnt **"How to find basis of a vector space?"** Denote basis of a vector sapce V by B. **If the set B conatins n vectors, then the coordinate system will make V act like $\mathbb{R}^n$.**

The following theorem states about the existence of coordinate system.

Theorem 1.9. *(**The unique representation theorem**)*

*Let $B = \{b_1, b_2, \cdots b_n\}$ be a basis of a vector space V. Then for each $v \in V, \exists$ a **unique** set of scalars $c_1, c_2, \cdots c_n$ such that*

$$v = c_1 b_1 + c_2 b_2 + \cdots + c_n b_n.$$

Proof: Let if possible $v \in V$ is written as follows

$$v = d_1 b_1 + d_2 b_2 + \cdots + d_n b_n$$

along with the v represented as

$$v = c_1 b_1 + c_2 b_2 + \cdots + c_n b_n$$

then $0 = v - v = (c_1 - d_1)b_1 + (c_2 - d_2)b_2 + \cdots + (c_n - d_n)b_n$.

This shows that $c_1 = d_1,\ c_2 = d_2 \cdots c_n = d_n$.

This proves the uniqueness of the scalars $c_1, c_2, \cdots c_n$.

Definition 1.11. *Let V be a vector space. Let $B = \{b_1, b_2, \cdots , b_n\}$ be a basis of V. Let $x \in V$. Then the **coordinates of x relative to the basis B (or the B-coordinates of x)** are the weight (scalars) $c_1, c_2, \cdots , c_n$ such that $x = c_1 b_1 + c_2 b_2 + \cdots + c_n b_n$.*

Notation: If $c_1, c_2, \cdots c_n$ are the B-coordinates of $x \in V$ then the vector in $\mathbb{R}^n$

$$[x]_B = \begin{bmatrix} c_1 \\ c_2 \\ \vdots \\ c_n \end{bmatrix}$$

is the coordinate vector of x (relative to B). This establishes the mapping $V \to \mathbb{R}^n$ defined as $x \mapsto [x]_B$. This is called **the coordinate mapping** (determined by B).

Example 1.57. *Let $\left\{ \begin{bmatrix} 1 \\ 0 \end{bmatrix}, \begin{bmatrix} 1 \\ 2 \end{bmatrix} \right\}$ be a basis of $\mathbb{R}^2$.*

Solution: Let $x \in \mathbb{R}^2$ with coordinate vector $[x]_B = \begin{bmatrix} -2 \\ 3 \end{bmatrix}$.

Then

$$= (\ 2) \begin{bmatrix} 1 \end{bmatrix} \quad \begin{bmatrix} 1 \end{bmatrix} \quad \begin{bmatrix} -2 \end{bmatrix} \quad \begin{bmatrix} 3 \end{bmatrix} \quad \begin{bmatrix} 1 \end{bmatrix}$$

This example illustrates the method of finding a vector whose coordinate vector and basis of the cooresponding vector space is given.

Example 1.58. *If $x \in \mathbb{R}^2$ is given as $\begin{bmatrix} 1 \\ 6 \end{bmatrix}$ with respect to the standard basis of $\mathbb{R}^2$ then the coordinate vector of x is given by*

$$x = 1 \begin{bmatrix} 1 \\ 0 \end{bmatrix} + 6 \begin{bmatrix} 0 \\ 1 \end{bmatrix} = 1e_1 + 6\,e_2$$

Solution: Here $[x]_B = \begin{bmatrix} 1 \\ 6 \end{bmatrix}$. Therefore, in general if the given basis of $\mathbb{R}^n$ is the standard basis then the coordinate vector of $x \in \mathbb{R}^n$ is itself.

Graphically the coordinate vector $[x]_B$ of a given vector $x \in \mathbb{R}^2$ with respect to basis B represents the position of x on B-graph paper as shown in the following figures.

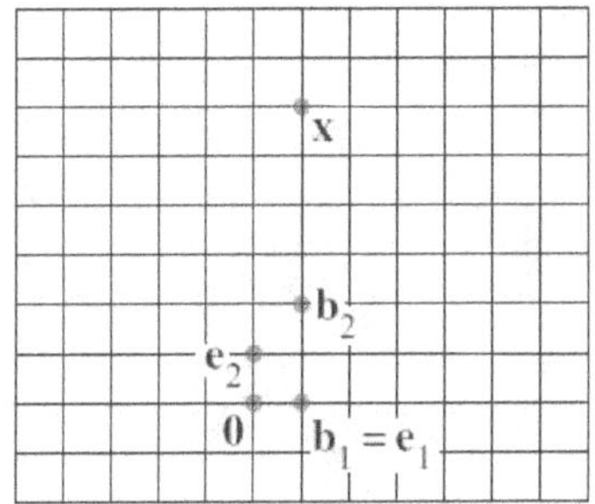
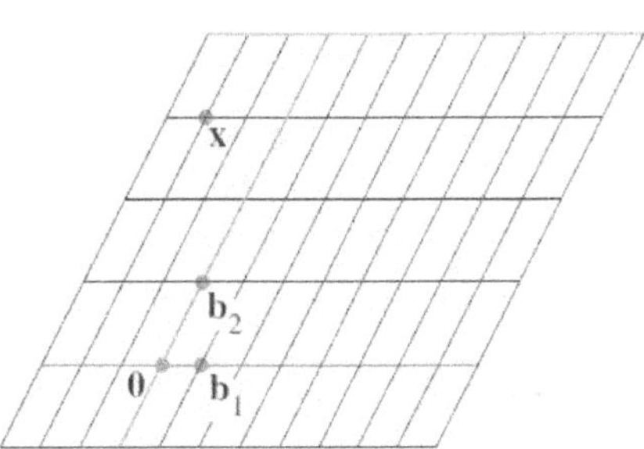

standard basis $\{e_1, e_2\}$ basis $B = \{b_1, b_2\}$.

Example 1.59. *Let $b_1 = \begin{bmatrix} 1 \\ -3 \end{bmatrix}$, $b_2 = \begin{bmatrix} 2 \\ -5 \end{bmatrix}$, $x = \begin{bmatrix} -2 \\ 1 \end{bmatrix}$, from the given information find $[x]_B$.*

Solution: We write the corresponding vector equation as,

$$c_1 b_1 + c_2 b_2 = x$$

This is,

$$c_1 \begin{bmatrix} 1 \\ -3 \end{bmatrix} + c_2 \begin{bmatrix} 2 \\ -5 \end{bmatrix} = \begin{bmatrix} -2 \\ 1 \end{bmatrix}$$

Solving the augmented matrix of the above linear system so formed we get

$$\begin{bmatrix} 1 & 2 & -2 \\ -3 & -5 & 1 \end{bmatrix} \sim \begin{bmatrix} 1 & 2 & -2 \\ 0 & 1 & -5 \end{bmatrix}$$

Rewriting the equation from the above echelon form we get $c_1 = 8$ and $c_2 = -5$.

Hence the coordinate vector $[x]_B = \begin{bmatrix} 8 \\ -5 \end{bmatrix}$.

This vector $[x]_B$ determines the position of vector x on the graph drawn with respect to basis $B = \{b_1, b_2\}$.

Example 1.60. *Given $B = \{b_1, b_2\}$ where $b_1 = \begin{bmatrix} 3 \\ 6 \\ 2 \end{bmatrix}$ and $b_2 = \begin{bmatrix} -1 \\ 0 \\ 1 \end{bmatrix}$. Let $x = \begin{bmatrix} 3 \\ 12 \\ 7 \end{bmatrix}$.*

Solution: We determine if $x \in$ Span $\{b_1, b_2\}$ and if so we find $[x]_B$.

To show that $x \in$ Span $\{b_1, b_2\}$ we solve the following vector equation

$$c_1 b_1 + c_2 b_2 = x$$

by row reducing the augmented matrix.

$$\begin{bmatrix} 3 & -1 & 3 \\ 6 & 0 & 12 \\ 2 & 1 & 7 \end{bmatrix} \sim \begin{bmatrix} 1 & 0 & 2 \\ 0 & 1 & 3 \\ 0 & 0 & 0 \end{bmatrix}$$

This shows that the corresponding linear system is consistent, hence $x \in$ span $\{b_1, b_2\}$ with $c_1 = 2, c_2 = 3$. Therefore, $[x]_B = \begin{bmatrix} 2 \\ 3 \end{bmatrix}$

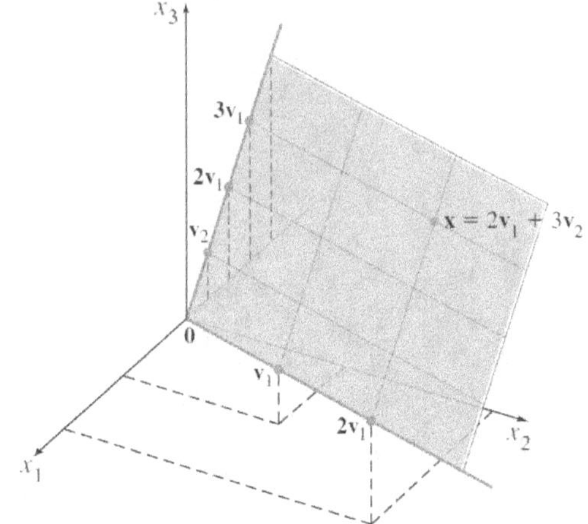

The above figure graphically shows the position of vector x with respect to basis $B = \{b_1, b_2\}$.

Example 1.61. *Let $B = \{b_1, b_2, b_3\} = \{1+t,\ 1+t^2,\ t+t^2\}$ be a basis for $\mathbb{P}_2$. Find coordinate vector of $p(t) = 6 + 3t - t^2$ relative to B.*

Solution: We write the vector equation

$$c_1 b_1 + c_2 b_2 + c_3 b_3 = p(t)$$

$$c_1(1+t) + c_2(1+t^2) + c_3(t+t^2) = 6 + 3t - t^2$$

$$\therefore (c_1 + c_2) + t(c_1 + c_3) + t^2(c_2 + c_3) = 6 + 3t - t^2$$

Equating the coefficient on both sides we get a linear system as follows

$$c_1 + c_2 = 6$$

$$c_1 + c_3 = 3$$

$$c_2 + c_3 = -1$$

Solving the above system we find $c_1 = 5, c_2 = 1$ and $c_3 = -2$

Hence $[p]_B = \begin{bmatrix} 5 \\ 1 \\ \end{bmatrix}$ is the required coordinate vector of $p(t)$.

Given a vector $x \in \mathbb{R}^n$, the entries in this vector x are the coordinates of x relative to the standard basis $\{e_1, e_2, \cdots e_n\}$ of $\mathbb{R}^n$. The change of coordinates of x in $\mathbb{R}^n$ with respect to another basis $B = \{b_1, b_2 \cdots b_n\}$ is carried out in the following way:

Let P_B denote the matrix whose columns are the basis vectors $b_1, b_2 \cdots b_n$.

$$P_B = [b_1, b_2 \cdots b_n]$$

Matrix P_B is known as **change of coordinate matrix.**

Then the vector equation

$$x = c_1 b_1 + c_2 b_2 + \cdots + c_n b_n$$

is equivalent to the matrix equation

$$x = P_B [x]_B \tag{1.3}$$

P_B matrix is a change of coordinate matrix from B to the standard matrix in $\mathbb{R}^n$.

The equation (A) is known as "**The change of coordinates equation**".

We known that the columns of matrix P_B are basis vectors of $\mathbb{R}^n$. So P_B is an invertible matrix. Hence

$$P_B^{-1} x = P_B^{-1} P_B [x]_B$$

$$\therefore P_B^{-1} x = [x]_B \tag{1.4}$$

The equation (1.3) and (1.4) are important for solving problems. When basis B and $[x]_B$ is given we use equation (1.3) to find vector $x \in \mathbb{R}^n$ with respect to standard basis.

On the other hand, when basis B and the vector $x \in \mathbb{R}^n$ with standard basis is given we find $[x]_B$ using equation (1.4).

The correspondence $x \mapsto [x]_B$ explained above produced by P_B^{-1} is known as **Coordinate Mapping.** It is a one-to-one linear transformation from $\mathbb{R}^n$ onto $\mathbb{R}^n$.

The following theorem generalizes the above discussion for any vector space V.

Theorem 1.10. *Let $B = \{b_1, b_2, \cdots b_n\}$ be a basis for a vector space V. Then the coordinate mapping $x \mapsto [x]_B$ is a one-to-one linear transformation from V onto $\mathbb{R}^n$.*

Example 1.62. *Let $B = \{1 + t^2, t + t^2, 1 + 2t + t^2\}$ be a basis of $\mathbb{P}_2$. Find the coordinate vector of $p(t) = 1 + 4t + 7t^2$ relative to B.*

Solution: We know that the standard basis of the space $\mathbb{P}_2$ of polynomials is $\{1, t, t^2\}$.

The given polynomial $p(t)$ is a linear combination of the standard basis vectors. To find $[p(t)]_B$ we use the following vector equation.

$$c_1 b_1 + c_2 b_2 + c_3 b_3 = p(t)$$

We find c_1, c_2, c_3.

$$c_1(1 + t^2) + c_2(t + t^2) + c_3(1 + 2t + t^2) = 1 + 4t + 7t^2$$

$$\therefore (c_1 + c_3) + t(c_2 + 2c_3) + t^2(c_1 + c_2 + c_3) = 1 + 4t + 7t^2$$

$$\begin{aligned} \therefore c_1 + c_3 &= 1 \\ c_2 + 2c_3 &= 4 \\ c_1 + c_2 + c_3 &= 7 \end{aligned}$$

Solving above system, we get $c_1 = 2, c_2 = 6, c_3 = -1$

Hence, $p(t) \mapsto [p(t)]_B$ is the coordinate map, where

$$[p(t)]_B = 2\,b_1 + 6b_2 + (-1)b_3$$

$$= [2 \quad 6 \quad -1]_B$$

This is the required coordinate vector.

Example 1.63. *For* $B = \left\{ \begin{bmatrix} 3 \\ -5 \end{bmatrix}, \begin{bmatrix} -4 \\ 6 \end{bmatrix} \right\}, \ x = \begin{bmatrix} 2 \\ -6 \end{bmatrix}, \ find\ [x]_B.$

Solution: We know that $P_B = \begin{bmatrix} 3 & -4 \\ -5 & 6 \end{bmatrix}$

$$\therefore P_B^{-1} = \frac{1}{-2}\begin{bmatrix} 6 & 4 \\ 5 & 3 \end{bmatrix}$$

$\therefore$ From equation (1.3) we get

$$P_B^{-1}x = [x]_B$$

$$\therefore \frac{1}{-2}\begin{bmatrix} 6 & 4 \\ 5 & 3 \end{bmatrix}\begin{bmatrix} 2 \\ -6 \end{bmatrix} = \frac{-1}{2}\begin{bmatrix} -12 \\ -8 \end{bmatrix} = \begin{bmatrix} 6 \\ 4 \end{bmatrix} = [x]_B$$

Example 1.64. *Given set* $S = \{1 + t^2, 4 + t + 5t^2, 3 + 2t\}$ *of polynomial in* $\mathbb{P}_2$*. Show that S is linearly dependent set using coordinate vectors.*

Solution: The coordinate mapping produces the coordinate vectors $(1, 0, 2), (4, 1, 5)$ and $(3, 2, 0)$ respectively. Write these vectors as columns of matrix A as, (for the system $Ax = O$)

$$\begin{bmatrix} 1 & 4 & 3 & 0 \\ 0 & 1 & 2 & 0 \\ 2 & 5 & 0 & 0 \end{bmatrix} \sim \begin{bmatrix} 1 & 4 & 3 & 0 \\ 0 & 1 & 2 & 0 \\ 0 & 0 & 0 & 0 \end{bmatrix}$$

The row reduced matrix has only two leading elements. This shows that the columns of A are linearly dependent. So the corresponding polynomials are linearly dependent.

1.6   The dimension of a vector space

In section 1.4, we have stated a bijection between a vector space V (with a basis B conataining n

This number 'n' denotes the dimension of the space V that is independent of the choice of the basis B. Also in Theorem 1.10 we have seen that a set of vectors in V conataining more than n **vectors** is linearly dependent.

That is, in the above situation each linearly independent set in space V has no more than n vectors.

Now, the following theorem states the exact number of vectors in a basis of a vector space V.

Theorem 1.11. *If a vector space V has a basis of n vectors, then every basis of V must consist of exactly n vectors.*

Proof: Let B_1 denote a basis of n vectors. Let B_2 be any other basis of V. B_2 is linearly independent by assumption. By theorem $1 \cdot 6$, B_2 has **no more than** n vectors. Similarly, B_2 is a basis and B_1 is a linearly independent set, B_2 has **at least** n **vectors**. Hence B_2 has exactly n vectors.

Definition 1.12. Dimension of a vector space V

The dimension of a vector space V is the number of vectors in a basis for V and it is denoted by $dim V$

A vector space V is said to be **finite-dimensional** if the vector space V is spanned by a finite set. That is, if the basis of V consist of a finite number of vectors then V is a finite dimensional vector space.

The dimension of the zero vector space $\{0\}$ is defined to be **zero**. If V is **not** spanned by a finite set, then V is said to be **infinite dimensional** vector space.

Example 1.65. *The following table indicates the standard bases and dimensions of some vector spaces.*

Solution: In the following table we have listed a few vector spaces with their respective standard bases.

Vector Space	Standard Basis	Dimension
$\mathbb{R}^n$	$\{e_1, e_2, \cdots, e_n\}$	n
$\mathbb{P}_2$	$\{1, t, t^2\}$	3
$\mathbb{P}_n$	$\{1, t, t^2, \cdots, t^n\}$	$n+1$
$\{0\}$	-	0
$\mathbb{P}$ (The space of all polynomials)	$\{1, t, t^2, \cdots\}$	infinite
$M_{2\times 2}$ (The set of all 2×2 matrices with real entries)	$\{M_1, M_2, M_3, M_4\}$ $M_1 = \begin{bmatrix} 1 & 0 \\ 0 & 0 \end{bmatrix}, M_2 = \begin{bmatrix} 0 & 1 \\ 0 & 0 \end{bmatrix},$ $M_3 = \begin{bmatrix} 0 & 0 \\ 1 & 0 \end{bmatrix}, M_4 = \begin{bmatrix} 0 & 0 \\ 0 & 1 \end{bmatrix},$	4

But, besides these mentioned bases a vector space can have a non-standard basis. For example, space $\mathbb{R}^3$ has a non-standard basis $S = \{(1, 2, -3), (1, -3, 2), (2, -1, 5)\}$. This shows that the basis for a vector space need not be unique. However the number of vectors in both the sets (the standard and

Theorem 1.12. *The basis theorem*

Let V be a vector space of dimension $p, p \geq 1$. Any linearly independent set of exactly 'p' elements in V is automatically a basis of V. Any set of exactly 'p' elements that spans V is automatically a basis of V.

We have studied **Nul** A and **Col** A of a given matrix A. We know that the pivot columns of matrix A form a basis of **Col** A, So we know the dimension of Col A as soon as we know the pivot columns of A.

On the other hand, finding a basis of **Nul** A needs more work. Suppose A is a given matrix of order $m \times n$. Let the matrix equation $Ax = O$ has k free variables. We have seen the standard method of finding a spanning set for Nul A. This produces exactly k linearly independent vectors say $u_1, u_2 \cdots u_k$ (one for each free variables). Hence the set $\{u_1, u_2 \cdots u_k\}$ is a basis of **Nul** A. **The number of free variables** determines the size of the basis.

From the above discussion we conclude the following:

The dimension of Nul A is the number of free variables in the equation $Ax = O$.

The dimension of Col A is the number of pivot columns of A.

Example 1.66. *Find the dimensions of the* **Nul** A *and* **Col** A *of the matrix A where*

$$A = \begin{bmatrix} -3 & 6 & -1 & 1 & -7 \\ 1 & -2 & 2 & 3 & -1 \\ 2 & -4 & 5 & 8 & -4 \end{bmatrix}$$

Solution: By row reduction of augmented matrix $[A \ O]$ we get an echelon form as follows:

$$\begin{bmatrix} 1 & -2 & 2 & 3 & -1 & 0 \\ 0 & 0 & 1 & 2 & -2 & 0 \\ 0 & 0 & 0 & 0 & 0 & 0 \end{bmatrix}$$

There are two leading variables say x_1, x_3 and three **free variables** say x_2, x_4, x_5.

Hence dim **Nul** $A = 3$. Also there are two pivot columns in the corresponding echelon form. So dim **Col** $A = 2$

Example 1.67. *Given set S of vectors as follows $S = \left\{ \begin{bmatrix} 1 \\ 0 \\ 2 \end{bmatrix}, \begin{bmatrix} 3 \\ 1 \\ 1 \end{bmatrix}, \begin{bmatrix} 9 \\ 4 \\ -2 \end{bmatrix}, \begin{bmatrix} -7 \\ -3 \\ 1 \end{bmatrix} \right\}$. Find dimension of the subspace spanned by these vectors.*

Solution: That is we find dimension of the subspace spanned by the columns of the following matrix.

$$A = \begin{bmatrix} 1 & 3 & 9 & -7 \\ 0 & 1 & 4 & -3 \\ 2 & 1 & -2 & 1 \end{bmatrix}$$

By row reducing matrix A we get the echelon form of the augmented matrix is

$$\begin{bmatrix} 1 & 3 & 9 & -7 & 0 \\ 0 & 1 & 4 & -3 & 0 \\ 0 & 0 & 0 & 0 & 0 \end{bmatrix}$$

We see that there are two pivot columns of the matrix hence the dimension of the subspace spanned by the given vectors is 2.

- **Rank**

we have studied Nul A and Col A of a given matrix A. Now we see space called **row space** associated with matrix A. **The set of all linear combinations of the row vectors is called the row space of matrix** A. Note that each row has 'n' entries in an $m \times n$ matrix. Hence row space of A i.e Row A is a subspace of $\mathbb{R}^n$. We know that rows of A are columns of A^T hence **Row** A is same as **Col** A^T.

The main task now is "**How to find the basis of the row space of** A?" The following theorem gives the answer to this question.

Theorem 1.13. *If two matrices A and B are row equivalent, then their row spaces are the same. If matrix B is in echelon form, then the non-zero rows form a basis for Row A as well as for B.*

Example 1.68. *For the following matrix A we find the row space, column space and null space of the matrix.*

$$A = \begin{bmatrix} -2 & -5 & 8 & 0 & -17 \\ 1 & 3 & -5 & 1 & 5 \\ 3 & 11 & -19 & 7 & 1 \\ 1 & 7 & -13 & 5 & -3 \end{bmatrix}$$

Solution: First we find row echelon form of A as

$$B = \begin{bmatrix} 1 & 3 & -5 & 1 & 5 \\ 0 & 1 & -2 & 2 & -7 \\ 0 & 0 & 0 & -4 & 20 \\ 0 & 0 & 0 & 0 & 0 \end{bmatrix}$$

By theorem (1.13) the first three rows form a basis of Row A and Row B. There are three pivot elements in first, second and fourth columns respectively of matrix B.

So the corresponding vectors of matrix A a

$$\begin{bmatrix} -2 \\ 1 \\ 3 \\ 1 \end{bmatrix}, \begin{bmatrix} -5 \\ 3 \\ 11 \\ 7 \end{bmatrix}, \begin{bmatrix} 0 \\ 1 \\ 7 \\ 5 \end{bmatrix}$$

form basis of Col A.

Now to find a basis of Nul A we find $RREF$ of A as follows

$$C = \begin{bmatrix} 1 & 0 & 1 & 0 & 1 \\ 0 & 1 & -2 & 0 & 3 \\ 0 & 0 & 0 & 1 & -5 \end{bmatrix}$$

Then the system $Cx = O$ becomes

$$x_1 + x_3 + x_5 = 0$$

$$x_2 - 2x_3 + 3x_5 = 0$$

$$x_4 - 5x_5 = 0$$

$$\therefore x_1 = -x_3 - x_5$$

$$x_2 = 2x_3 - 3x_5$$

$$x_4 = 5x_5$$

$$\therefore \begin{bmatrix} x_1 \\ x_2 \\ x_3 \\ x_4 \\ x_5 \end{bmatrix} = \begin{bmatrix} -x_3 - x_5 \\ 2x_3 - 3x_5 \\ x_3 \\ 5x_5 \\ x_5 \end{bmatrix} = x_3 \begin{bmatrix} -1 \\ 2 \\ 1 \\ 0 \\ 0 \end{bmatrix} + x_5 \begin{bmatrix} -1 \\ -3 \\ 0 \\ 5 \\ 1 \end{bmatrix}$$

$$\underset{v_1}{\uparrow} \qquad \underset{v_2}{\uparrow}$$

$\therefore$ The basis of Nul $A = \{v_1, v_2\}$.

Rank $A = $ The number of vectors in basis of Col A

$= 3$

Definition 1.13. : Rank

The rank of a matrix A is the dimension of the Col A,

We know that Row $A = $ Col A^T, the dimension of row sapce of A (Row A) is rank of A^T.

The dimension of Nul A is known as **Nullity** A.

Theorem 1.14. :(The rank Theorem)

The dimensions of the columns space (Col A) and rowspace (Row A) of an $m \times n$ matrix A are equal. This common dimension that is Rank A also equals the number of pivot positions of A and satisfies the following equation.

$$\text{Rank } A + \text{ Nullity } A = n$$

Theorem 1.15. *Let A be an $m \times n$ matrix. Then the following statements are each equivalent to the statement that A is an invertible matrix.*

(a) The columns of A form a basis of $\mathbb{R}^n$.

(b) Col $A = \mathbb{R}^n$.

(c) dim Col $A = n$

(d) rank $A = n$

(e) Nul A = {0}

(f) dim *Nul A = 0.*

We have already studied 12 equivalent statemennts of **Invertible Matrix Theorem** in the Matrix Algebra course in the last term. Theorem 1.15 above extent it further.

Example 1.69. *For a given matrix A below, matrix B given is such that $A \sim B$. Without calculation we here by list rank A and* dim *Nul A. then we find bases for Col A, Row A and Nul A.*

$$A = \begin{bmatrix} 2 & -3 & 6 & 2 & 5 \\ -2 & 3 & -3 & -3 & -4 \\ 4 & -6 & 9 & 5 & 9 \\ -2 & 3 & 3 & -4 & 1 \end{bmatrix}, \quad B = \begin{bmatrix} 2 & -3 & 6 & 2 & 5 \\ 0 & 0 & 3 & -1 & 1 \\ 0 & 0 & 0 & 1 & 3 \\ 0 & 0 & 0 & 0 & 0 \end{bmatrix}$$

Solution: From matrix B it is clear that there are three non-zero rows in it. So Rank $A = 3$. From **Rank-Nullity theorem** we get dim Nul $A = 5 - 3 = 2$.

Basis for Col A is

$$\left\{ \begin{bmatrix} 2 \\ -2 \\ 4 \\ 2 \end{bmatrix}, \begin{bmatrix} 6 \\ -3 \\ 9 \\ 3 \end{bmatrix}, \begin{bmatrix} 2 \\ -3 \\ 5 \\ -4 \end{bmatrix} \right\}$$

Basis for Row $A = \{(2, -3, 6, 2, 5),\ (-2, 3, -3, -3, -4),\ (4, -6, 9, 5, 9)\}$ To find basis of Nul A, From matrix B we write the following linear system,

$$2x_1 - 3x_2 + 6x_3 + 2x_4 + 5x_5 = 0$$

$$3x_3 - x_4 + x_5 = 0$$

$$x_4 + 3x_5 = 0$$

Here, we see that the variables x_1, x_3 and x_4 are leading variables. x_2, x_5 are free variables. So,

$$x_1 = \frac{3}{2}x_2 + \frac{9}{2}x_5$$

$$x_2, x_5 \text{ are free}$$

$$x_3 = \frac{-4}{3}x_5$$

$$x_4 = -3x_5$$

Hence

$$\begin{bmatrix} x_1 \\ x_2 \\ x_3 \\ x_4 \\ x_5 \end{bmatrix} = x_2 \begin{bmatrix} \frac{3}{2} \\ 1 \\ 0 \\ 0 \\ 0 \end{bmatrix} + x_5 \begin{bmatrix} \frac{9}{2} \\ 0 \\ \frac{-4}{3} \\ -3 \\ 1 \end{bmatrix}$$

$$\uparrow \qquad\qquad \uparrow$$

$\therefore$ Basis of Nul $A = \{v_1, v_2\}$

EXERCISE

1. Test whether the following sets are vector spaces with respect to the standard operations of vector addition and scalar multiplication unless and otherwise stated.

   (a) Set V is the first quadrant in XY-plane. That is, $V = \left\{ \begin{bmatrix} x \\ y \end{bmatrix} / x \geq 0, y \geq 0 \right\}$

   (b) Set W be the union of first and third quadrant in the XY-plane. That is, $W = \left\{ \begin{bmatrix} x \\ y \end{bmatrix} / xy \geq 0 \right\}$

   (c) H is a point inside and on the unit circle in the XY-plane. That is, $H = \left\{ \begin{bmatrix} x \\ y \end{bmatrix} / x^2 + y^2 \leq 1 \right\}$

   (d) $S = \{(x, y) / x, y \in \mathbb{R}\}$ for set S vector addition and scalar multiplication is defined as follows:
   $(x_1, y_1) + (x_2, y_2) = (|x_1 + x_2|, |y_1 + y_2|)$
   $k(x, y) = (|kx|, |ky|)$.

2. Test whether the following subsets are subspaces of the respective vector spaces or not. The operations of vector addition and scalar multiplication are standard unless and otherwise stated.

   (a) $V = \mathbb{P}_n$, $S = \{p(t) / p(t) = at^2, a \in \mathbb{R}\}$

   (b) $V = \mathbb{P}_n$, $S = \{p(t) / p(t) = a + t^2, a \in \mathbb{R}\}$

   (c) $V = \mathbb{P}_n$, $H = \{p(t) / p(0) = 0\}$

   (d) $V = \mathbb{R}^3$, $T = \left\{ \begin{bmatrix} s \\ 3s \\ 2s \end{bmatrix} / s \in \mathbb{R} \right\}$ Find a vector $v \in \mathbb{R}^3$ such that $T = $ span $\{v\}$.

   (e) $V = \mathbb{R}^3$, $H = \left\{ \begin{bmatrix} 2t \\ 0 \\ -t \end{bmatrix} / t \in \mathbb{R} \right\}$

   (f) $W = \left\{ \begin{bmatrix} s + 3t \\ s - t \\ 2s - t \\ 4t \end{bmatrix} / s, t \in \mathbb{R} \right\}$, $V = \mathbb{R}^4$

   (g) $V = \mathbb{R}^3$, $T = \left\{ \begin{bmatrix} 3a + b \\ 4 \\ a - 5b \end{bmatrix} / a, b \in \mathbb{R} \right\}$

   (h) $V = M_{2 \times 2}$, $S = \left\{ \begin{bmatrix} a & b \\ 0 & d \end{bmatrix} / a, b, d \in \mathbb{R} \right\}$

   (i) $V = C[a, b] = A$ vector space of real valued continuous functions on $[a, b]$.
   $S = \{f / f \in C[a, b], f(a) = f(b)\}$.

3. Determine if vector y is in the subspace of $\mathbb{R}^4$ spanned by the columns of matrix A, where

$$y = \begin{bmatrix} -4 \\ -8 \\ 6 \\ -5 \end{bmatrix}, \quad A = \begin{bmatrix} 3 & -5 & -9 \\ 8 & 7 & -6 \\ -5 & -8 & 3 \\ 2 & -2 & -9 \end{bmatrix}$$

4. Determine if $W = \begin{bmatrix} 5 \\ -3 \\ 2 \end{bmatrix}$ is in Nul A, where $A = \begin{bmatrix} 5 & 21 & 19 \\ 13 & 23 & 2 \\ 8 & 14 & 1 \end{bmatrix}$.

5. Find an explicit description of Nul A by listing vectors that span Nul A.

(a) $\begin{bmatrix} 1 & 3 & 5 & 0 \\ 0 & 1 & 4 & -2 \end{bmatrix}$ (b) $\begin{bmatrix} 1 & -2 & 0 & 4 & 0 \\ 0 & 0 & 1 & -9 & 0 \\ 0 & 0 & 0 & 0 & 1 \end{bmatrix}$

6. Find k in each of the following such that Nul A is a subspace of $\mathbb{R}^k$.

(a) $A = \begin{bmatrix} 7 & -2 & 0 \\ -2 & 0 & -5 \\ 0 & -5 & 7 \\ -5 & 7 & -2 \end{bmatrix}$ (b) $A = \begin{bmatrix} 4 & 5 & -2 & 6 & 0 \\ 1 & 1 & 0 & 1 & 0 \end{bmatrix}$ (c) $A = \begin{bmatrix} 1 & -3 & 9 & 0 & -5 \end{bmatrix}$

7. Let $A = \begin{bmatrix} -6 & 12 \\ -3 & 6 \end{bmatrix}$ and $W = \begin{bmatrix} 2 \\ 1 \end{bmatrix}$. Determine if w is in Col A and / or Nul A.

8. Find matrix A in each of the following such that the given set W is Col A.

(a) $W = \left\{ \begin{bmatrix} 2s + 3t \\ r + s - 2t \\ 4r + s \\ 3r - s - t \end{bmatrix} / r, s, t \in \mathbb{R} \right\}$ (b) $W = \left\{ \begin{bmatrix} 6a - b \\ a + b \\ -7a \end{bmatrix} / a, b \in \mathbb{R} \right\}$

9. In the following sets W, show that W is not a vector space.

(a) $W = \left\{ \begin{bmatrix} a \\ b \\ c \end{bmatrix} / a + b + c = 2 \right\}$ (b) $W = \left\{ \begin{bmatrix} b - 5d \\ 2b \\ 2d + 1 \\ d \end{bmatrix} / b, d \in \mathbb{R} \right\}$

(c) $W = \left\{ \begin{bmatrix} b - 2d \\ 5 + d \\ b + 3d \\ d \end{bmatrix} / b, d \in \mathbb{R} \right\}$

10. Let $A = \begin{bmatrix} -6 & 12 \\ -3 & 6 \end{bmatrix}, \quad w = \begin{bmatrix} 2 \\ 1 \end{bmatrix}$. Determine if w is in Col A. Is w in Nul A?

11. State **true** or **false.** A denotes an $m \times n$ matrix. Justify your answer.

(a) The set $S = \left\{ \begin{bmatrix} 1 & b \\ c & d \end{bmatrix} / b, c, d \in \mathbb{R} \right\}$ is a subspace of $M_{2\times 2}$.

(b) The set $T = \{(1, 1), (0, 1)\}$ is linearly independent subset of $\mathbb{R}^2$.

(c) A null space is a vector space.

(e) If the equation $Ax = b$ is consistent then Col A is $\mathbb{R}^m$.

(f) Col A is the range of the mapping $x \longmapsto Ax$.

12. Determine whether w is in the column space of A, the Nul A or both.

$$\text{Where } w = \begin{bmatrix} 1 \\ 1 \\ -1 \\ -3 \end{bmatrix}, \quad A = \begin{bmatrix} 7 & 6 & -4 & 1 \\ -5 & -1 & 0 & -2 \\ 9 & -11 & 7 & -3 \\ 19 & -9 & 7 & 1 \end{bmatrix}$$

13. Let $T : V \to W$ be a linear transformation. (V and W are vector spaces) prove that range (T) is a subspace of W.

14. Let $M_{2 \times 2}$ be the vector space of all 2×2 matrices, define $T : M_{2 \times 2} \to M_{2 \times 2}$ by $T(A) = A + A^T$ where $A = \begin{bmatrix} a & b \\ c & d \end{bmatrix}$. Show that T is a linear transformation; it is an onto map. Find $\ker(T)$.

15. Show that the mapping $T : M_{2 \times 2} \to \mathbb{P}_1$ defined by $T\left(\begin{bmatrix} a & b \\ c & d \end{bmatrix} \right) = (a - c) + (2b + d)x$ is a linear transformation. Find $\ker(T)$.

16. Consider the map $T : \mathbb{P}_2 \to \mathbb{R}^2$ defined by $T(a_0 + a_1 x + a_2 x^2) = (a_1 + 2a_1 + a_2, a_0 - a_2)$ is a linear transformation. Find $\ker(T)$ and range (T).

17. In each of the following determine if a mapping $T : M_{2 \times 2} \to \mathbb{R}^2$ is a linear transformation. If yes, find $\ker(T)$.

(a) $T\left(\begin{bmatrix} a & b \\ c & d \end{bmatrix} \right) = (ad + 1, b + c)$

(b) $T\left(\begin{bmatrix} a & b \\ c & d \end{bmatrix} \right) = (2a - b - d, 0)$

(c) $T\left(\begin{bmatrix} a & b \\ c & d \end{bmatrix} \right) = (a - b - 2c, b + c - d)$

(d) $T\left(\begin{bmatrix} a & b \\ c & d \end{bmatrix} \right) = \left(\det \begin{bmatrix} a & b \\ c & d \end{bmatrix}, 0 \right)$

(e) $T\left(\begin{bmatrix} a & b \\ c & d \end{bmatrix} \right) = (2a - bc + d, 2a + 3d)$

18. In each of the following map $T : \mathbb{R}^2 \to \mathbb{R}^3$. Determine if T is a linear transformation.

(a) $T(x_1, x_2) = (2x_1 + x_2, -7x_2, -x_1 - x_2)$

(b) $T(x_1, x_2) = (x_1 + x_2 + 1, -4x_1 + x_2, 2x_2)$

(c) $T(x_1, x_2) = (x_1 + x_2, 4x_1 - x_2, -x_1 + x_2)$

(d) $T(x_1, x_2) = (3x_1 + 2x_2, -2x_1 + x_2, 3x_1)$

(e) $T(x_1, x_2) = (x_1^2, x_2^2, x_1 x_2)$

19. Let $T : \mathbb{R}^2 \to \mathbb{R}^3$ be a linear transformation, defined as $T\left(\begin{bmatrix} x \\ y \end{bmatrix}\right) = \begin{bmatrix} y \\ -5x + 13y \\ -7x + 16y \end{bmatrix}$.

    Find $\ker(T)$.

20. Find standard matrix of transformation for $T : \mathbb{R}^3 \to \mathbb{R}^5$ in each of the following:

    (a) $T\left(\begin{bmatrix} x \\ y \\ z \end{bmatrix}\right) = \begin{bmatrix} 5x - z \\ 2y - 4x \\ y + z \\ -x - y \\ 5y \end{bmatrix}$.

    (b) $T\left(\begin{bmatrix} x_1 \\ x_2 \\ x_3 \end{bmatrix}\right) = \begin{bmatrix} 2x_1 - 3x_2 - 4x_3 \\ -x_1 - x_2 - 5x_3 \\ 4x_2 \\ 0 \\ 3x_1 + x_3 \end{bmatrix}$.

21. Determine which of the following sets of vectors form a basis of $\mathbb{R}^3$. Justify your answer.

    (a) $\begin{bmatrix} 1 \\ 0 \\ 0 \end{bmatrix}, \begin{bmatrix} 1 \\ 1 \\ 0 \end{bmatrix}, \begin{bmatrix} 1 \\ 1 \\ 1 \end{bmatrix}$

    (b) $\begin{bmatrix} 1 \\ 0 \\ -2 \end{bmatrix}, \begin{bmatrix} 3 \\ 2 \\ -4 \end{bmatrix}, \begin{bmatrix} -3 \\ -5 \\ 1 \end{bmatrix}$

    (c) $\begin{bmatrix} 1 \\ 2 \\ -3 \end{bmatrix}, \begin{bmatrix} -4 \\ -5 \\ 6 \end{bmatrix}$

    (d) $\begin{bmatrix} 1 \\ -4 \\ 3 \end{bmatrix}, \begin{bmatrix} 0 \\ 3 \\ -1 \end{bmatrix}, \begin{bmatrix} 3 \\ -5 \\ 4 \end{bmatrix}, \begin{bmatrix} 0 \\ 2 \\ -2 \end{bmatrix}$

22. Find a basis for the space spanned by the vectors v_1, v_2, v_3, v_4, v_5.

    (a) $\begin{bmatrix} 1 \\ 0 \\ -3 \\ 2 \end{bmatrix}, \begin{bmatrix} 0 \\ 1 \\ 2 \\ -3 \end{bmatrix}, \begin{bmatrix} -3 \\ -4 \\ 1 \\ 6 \end{bmatrix}, \begin{bmatrix} 1 \\ -3 \\ -8 \\ 7 \end{bmatrix}, \begin{bmatrix} 2 \\ 1 \\ -6 \\ 9 \end{bmatrix}$

    (b) $\begin{bmatrix} 1 \\ 0 \\ 0 \\ 1 \end{bmatrix}, \begin{bmatrix} -2 \\ 1 \\ -1 \\ 1 \end{bmatrix}, \begin{bmatrix} 6 \\ -1 \\ 2 \\ -1 \end{bmatrix}, \begin{bmatrix} 5 \\ -3 \\ 3 \\ -4 \end{bmatrix}, \begin{bmatrix} 0 \\ 3 \\ -1 \\ 1 \end{bmatrix}$

    (c) $\begin{bmatrix} 8 \\ 9 \\ -3 \\ -6 \\ 0 \end{bmatrix}, \begin{bmatrix} 4 \\ 5 \\ 1 \\ -4 \\ 4 \end{bmatrix}, \begin{bmatrix} -1 \\ -4 \\ -9 \\ 6 \\ -7 \end{bmatrix}, \begin{bmatrix} 6 \\ 8 \\ 4 \\ -7 \\ 10 \end{bmatrix}, \begin{bmatrix} -1 \\ 4 \\ 11 \\ -8 \\ -7 \end{bmatrix}$

23. Find a basis for the subspace spanned by the given set of vectors in the indicated vector space.

    (a) $S = \{(1, 3, 1, -2), (1, 4, 3, -1), (2, 3, -4, -7), (3, 8, 1, -7)\}$ in $\mathbb{R}^4$.

    (b) $S = \{(-1, 2, 0, 0, 3), (2, -5, -3, -2, 6), (0, 5, 15, 10, 0), (2, 6, 18, 8, 6)\}$ in $\mathbb{R}^5$.

24. Find basis for the null space of the matrix

    $$A = \begin{bmatrix} 1 & 0 & -5 & 1 & 4 \\ -2 & 1 & 6 & -2 & -2 \\ 0 & 2 & -8 & 1 & 9 \end{bmatrix}$$

25. Find basis of Col A in each of the following:

    (a) $A = \begin{bmatrix} -2 & 4 & -2 & -4 \\ 2 & -6 & -3 & 1 \end{bmatrix}$

    (b) $A = \begin{bmatrix} 1 & 2 & -5 & 11 & -3 \\ 2 & 4 & -5 & 15 & 2 \\ 1 & 2 & 0 & 4 & 5 \end{bmatrix}$

26. State **true** or **false**. Justify each answer.

    (a) If $H = \text{Span } \{b_1, b_2, \cdots, b_p\}$ then the set $\{b_1, b_2, \cdots, b_p\}$ is a basis for H.

    (b) The columns of an invertible $n \times n$ matrix form a basis for $\mathbb{R}^n$.

    (c) A basis is a spanning set that is as large as possible.

27. Find a basis for the set of vectors in $\mathbb{R}^2$ on the line $y = 5x$.

28. Let $v_1 = \begin{bmatrix} 7 \\ 4 \\ -9 \\ -5 \end{bmatrix} . v_2 = \begin{bmatrix} 4 \\ -7 \\ 2 \\ 5 \end{bmatrix}, v_3 = \begin{bmatrix} 1 \\ -5 \\ 3 \\ 4 \end{bmatrix}.$

    It can be verified that $v_1 - 3v_2 + 5v_3 = 0$. Use this information to find basis for space $H = \text{Span}\{v_1, v_2, v_3\}$.

29. Consider the polynomials $p_1(t) = 1 + t^2$, $p_2(t) = 1 - t^2$. Determine if $\{p_1, p_2\}$ is linearly independent set in $\mathbb{P}_3$. Justify.

30. Let $V = \mathbb{R}^3$, let $S = \{(1, 1, 1), (2, 1, 1), (3, 0, 0)\}$. show that S is a basis for V.

31. Show that the set $S = \{(0, 1, 2), (0, 2, 3), (0, 3, 1)\}$ in $\mathbb{R}^3$ is not a basis of $\mathbb{R}^3$.

32. Find a basis of the row space of A where

$$A = \begin{bmatrix} 1 & 2 & 3 & 0 \\ -1 & 1 & 1 & 4 \\ 3 & 6 & 9 & 0 \end{bmatrix}$$

33. Find a basis of the column space of A where

$$A = \begin{bmatrix} 2 & 3 & 5 & 7 & 4 \\ -1 & 2 & 1 & 0 & -2 \\ 4 & 1 & 5 & 9 & 8 \end{bmatrix}$$

34. Find a basis of the solution space for the following linear system

$$x + 3y + 2z = 0$$

$$x + 5y + z = 0$$

$$3x + 5y + 8z = 0$$

35. Find the vector x determined by the given coordinate vector $[x]_B$ and the given basis B.

    (a) $B = \left\{ \begin{bmatrix} 3 \\ -5 \end{bmatrix}, \begin{bmatrix} -4 \\ 6 \end{bmatrix} \right\}, [x]_B = \begin{bmatrix} 5 \\ 3 \end{bmatrix}$ (b) $B = \left\{ \begin{bmatrix} 4 \\ 5 \end{bmatrix}, \begin{bmatrix} 6 \\ 7 \end{bmatrix} \right\}, [x]_B = \begin{bmatrix} 8 \\ -5 \end{bmatrix}$

    (c) $B = \left\{ \begin{bmatrix} 1 \\ -4 \\ 3 \end{bmatrix}, \begin{bmatrix} 5 \\ 2 \\ -2 \end{bmatrix}, \begin{bmatrix} 4 \\ -7 \\ 0 \end{bmatrix} \right\}, [x]_B = \begin{bmatrix} 3 \\ 0 \\ -1 \end{bmatrix}$

    (d) $B = \left\{ \begin{bmatrix} -1 \\ 2 \end{bmatrix}, \begin{bmatrix} 3 \\ -5 \end{bmatrix}, \begin{bmatrix} 4 \\ -7 \end{bmatrix} \right\} [x]_B = \begin{bmatrix} -4 \\ 8 \end{bmatrix}$

36. Find the coordinate vector $[x]_B$ of vector x relative to the given basis $B = \{b_1, b_2, \cdots, b_n\}$

(a) $b_1 = \begin{bmatrix} 1 \\ -2 \end{bmatrix}$, $b_2 = \begin{bmatrix} 5 \\ -6 \end{bmatrix}$, $x = \begin{bmatrix} 4 \\ 0 \end{bmatrix}$

(b) $b_1 = \begin{bmatrix} 1 \\ -1 \\ -3 \end{bmatrix}$, $b_2 = \begin{bmatrix} -3 \\ 4 \\ 9 \end{bmatrix}$, $b_3 = \begin{bmatrix} 2 \\ -2 \\ 4 \end{bmatrix}$, $x = \begin{bmatrix} 8 \\ -9 \\ 6 \end{bmatrix}$

(c) $b_1 = \begin{bmatrix} 1 \\ 0 \\ 3 \end{bmatrix}$, $b_2 = \begin{bmatrix} 2 \\ 1 \\ 8 \end{bmatrix}$, $b_3 = \begin{bmatrix} 1 \\ -1 \\ 2 \end{bmatrix}$, $x = \begin{bmatrix} 3 \\ -5 \\ 4 \end{bmatrix}$

37. The set $B = \{1 - t^2, t - t^2, 2 - 2t + t^2\}$ is a basis for $\mathbb{P}_2$. Find the coordinate vector of $p(t) = 3 + t - 6t^2$ relative to B.

38. Find the change of coordinate matrix P_B from B to the standard basis in $\mathbb{R}^n$.

(a) $B = \left\{ \begin{bmatrix} 2 \\ -9 \end{bmatrix}, \begin{bmatrix} 1 \\ 8 \end{bmatrix} \right\}$ (b) $B = \left\{ \begin{bmatrix} 3 \\ -1 \\ 4 \end{bmatrix}, \begin{bmatrix} 2 \\ 0 \\ -5 \end{bmatrix}, \begin{bmatrix} 8 \\ -2 \\ 7 \end{bmatrix} \right\}$

39. Use an inverse matrix $\left(P_B^{-1}\right)$ to find $[x]_B$ for the given x and B.

$B = \left\{ \begin{bmatrix} 4 \\ 5 \end{bmatrix}, \begin{bmatrix} 6 \\ 7 \end{bmatrix} \right\}$, $x = \begin{bmatrix} 2 \\ 0 \end{bmatrix}$

40. The set $B = \{1+t, 1+t^2, t+t^2\}$ is a basis for $\mathbb{P}_2$. Find the coordinate vector of $p(t) = 6 + 3t - t^2$ relative to B.

41. Use coordinate vectors to show that the set $S = \{1+t^2, 1-3t^2, 1+t-3t^2\}$ is linearly independent in $\mathbb{P}_2$.

42. Find the dimension of the subspace spanned by the given vectors.

$$\begin{bmatrix} 1 \\ -2 \\ 0 \end{bmatrix}, \begin{bmatrix} -3 \\ 4 \\ 1 \end{bmatrix}, \begin{bmatrix} -8 \\ 6 \\ 5 \end{bmatrix}, \begin{bmatrix} -3 \\ 0 \\ 7 \end{bmatrix}$$

43. Determine the dimension of Nul A and Col A for the matrices shown below.

(a) $A = \begin{bmatrix} 1 & 0 & 9 & 5 \\ 0 & 0 & 1 & -4 \end{bmatrix}$ (b) $A = \begin{bmatrix} 1 & -6 & 9 & 0 & -2 \\ 0 & 1 & 2 & -4 & 5 \\ 0 & 0 & 0 & 5 & 1 \\ 0 & 0 & 0 & 0 & 0 \end{bmatrix}$

(c) $A = \begin{bmatrix} 1 & 3 & -4 & 2 & -1 & 6 \\ 0 & 0 & 1 & -3 & 7 & 0 \\ 0 & 0 & 0 & 1 & 4 & -3 \\ 0 & 0 & 0 & 0 & 0 & 0 \end{bmatrix}$ (d) $A = \begin{bmatrix} 1 & -1 & 0 \\ 0 & 4 & 7 \\ 0 & 0 & 5 \end{bmatrix}$

(e) $A = \begin{bmatrix} 1 & 4 & -1 \\ 0 & 7 & 0 \\ 0 & 0 & 0 \end{bmatrix}$

44. Show that the set of polynomials $S = \{1 \ 2t, \quad 2 + 4t^2 \quad 12t + 8t^3\}$ forms a basis of $\mathbb{P}_3$

45. State **True** or **False**. Justify your answer.

    (a) If $\dim V = p$ and if S is a linearly dependent subset of V, then S contains more than p vectors.

    (b) Let V be a vector space of finite dimension. If the set S spans V and if T is a subset of V that contains more vectors than S, then T is linearly dependent.

    (c) The number of pivot columns of a matrix equals the dimension of its column space.

    (d) The dimension of a vector space $\mathbb{P}_4$ is 4.

    (e) $\mathbb{R}^2$ is a two dimensional subspace of $\mathbb{R}^3$.

    (f) The number of variables in the equation $Ax = 0$ equals the dimension of the Nul A.

46. For each of the following matrices A find rank A, Nul A, bases for Col A, Row A and Nul A.

$$\text{(a) } A = \begin{bmatrix} 1 & -4 & 9 & -7 \\ -1 & 2 & -4 & 1 \\ 5 & -6 & 10 & 7 \end{bmatrix} \qquad \text{(b) } A = \begin{bmatrix} 1 & -3 & 4 & -1 & 9 \\ -2 & 6 & -6 & -1 & -10 \\ -3 & 9 & -6 & -6 & -3 \\ 3 & -9 & 4 & 9 & 0 \end{bmatrix}$$

$$\text{(c) } A = \begin{bmatrix} 1 & 1 & -3 & 7 & 9 & -9 \\ 1 & 2 & -4 & 10 & 13 & -12 \\ 1 & -1 & -1 & 1 & 1 & -3 \\ 1 & -3 & 1 & -5 & -7 & 3 \\ 1 & -2 & 0 & 0 & -5 & -4 \end{bmatrix} \qquad \text{(d) } A = \begin{bmatrix} 2 & -1 & 1 & -6 & 8 \\ 1 & -2 & -4 & 3 & -2 \\ -7 & 8 & 10 & 3 & -10 \\ 4 & -5 & -7 & 0 & 4 \end{bmatrix}$$

47. Answer the following:

    (a) If a 3×8 matrix A has rank 3, find dim Nul A, dim Row A, rank A^T.

    (b) If a 6×3 matrix A has rank 3, find dim Nul A, dim Row A, rank A^T.

    (c) Suppose a 4×7 matrix A has four pivot columns. Is Col $A = \mathbb{R}^4$? Is Nul $A = \mathbb{R}^3$? Explain your answers.

    (d) If the Nul A of a 5×6 matrix A is 4-dimensional, what is the dimension of the Col A?

SOLUTIONS

1   a) V is not a subspace, since for $u = \begin{bmatrix} 1 \\ 2 \end{bmatrix}$ and $c = -1$, $c.u \notin V$.

    b) W is not a subspace since for $u = \begin{bmatrix} 1 \\ 2 \end{bmatrix}$ and $v = \begin{bmatrix} -3 \\ -1 \end{bmatrix}$, $u + v = \begin{bmatrix} -2 \\ 1 \end{bmatrix} \notin W$

    c) H is not a subspace since for $u = \begin{bmatrix} 0.5 \\ 0.5 \end{bmatrix}$ and $c = 4$, $c.u = \begin{bmatrix} 2 \\ 2 \end{bmatrix} \notin H$.

    d) S is not a subspace, since for $k = 1$, $l = -1$ and $u = \begin{bmatrix} 1 \\ 1 \end{bmatrix}$, $(k+l)u \neq ku + lu$

2   a) S is a subspace since $S = span\{t^2\}$

b) S is not a subspace since the zero vector does not belong to S

c) Yes H is a subspace of V

d) H is a subspace since $H = span\{v\}$ where $v = \begin{bmatrix} 1 \\ 3 \\ 2 \end{bmatrix}$

e) H is a subspace since $H = span\{v\}$ where $v = \begin{bmatrix} 2 \\ 0 \\ 1 \end{bmatrix}$

f) W is a subspace since $H = span\{v_1, v_2\}$ where $v_1 = \begin{bmatrix} 1 \\ 1 \\ 2 \\ 0 \end{bmatrix}$ and $v_2 = \begin{bmatrix} 3 \\ -1 \\ -1 \\ 4 \end{bmatrix}$

g) T is not a subspace since zero vector does not belong to T.

h) Yes, S is a subspace of V.

i) Yes, S is a subspace of V.

3 Yes, y is in subspace spanned by columns

4 $w \in Nul A$

5   a) Nul A $= span\left\{ \begin{bmatrix} -6 \\ 2 \\ 0 \\ 1 \end{bmatrix}, \begin{bmatrix} 7 \\ -4 \\ 1 \\ 0 \end{bmatrix} \right\}$ Nul A $= span\left\{ \begin{bmatrix} 2 \\ 1 \\ 0 \\ 0 \\ 0 \end{bmatrix}, \begin{bmatrix} -4 \\ 0 \\ 9 \\ 1 \\ 0 \end{bmatrix} \right\}$

6 a) $k = 3$,   b) $k = 5$,   c)$k = 5$

7 w is in both Nul(A) and Col(A)

8 (a) $\begin{bmatrix} 2 & 0 & 3 \\ 1 & 1 & -2 \\ 1 & 4 & 0 \\ -1 & 3 & -1 \end{bmatrix}$

(b) $\begin{bmatrix} 6 & -1 \\ 1 & 1 \\ -7 & 0 \end{bmatrix}$

9 (a) Not a subspace since it does not have zero vector

(b) Not a subspace since it does not have zero vector

(c) Not a subspace since it does not have zero vector

10 w is in Co(A) as well as in Nul(A)

a) False b) True c) True d) False e) False f) True

12 w is in Col(A) but not in Nul(A)

13 zero vector is in Ran T and it satisfies the closure properties

14 Check that $T(A+B) = T(A) + T(B)$ and $T(kA) = kT(A)$

15 $Ker(T) = span\left\{ \begin{bmatrix} 1 \\ 0 \\ 1 \\ 0 \end{bmatrix}, \begin{bmatrix} 0 \\ \frac{-1}{2} \\ 0 \\ 1 \end{bmatrix} \right\}$

16 $Ker(T) = span\left\{ \begin{bmatrix} 1 \\ -1 \\ 1 \end{bmatrix} \right\}$, Range(T)$= span\left\{ \begin{bmatrix} 1 \\ 1 \end{bmatrix}, \begin{bmatrix} 2 \\ 0 \end{bmatrix} \right\}$

17 a) Not a linear transformation

   b) Yes, a linear transformation. $Ker(T) = span\left\{ \begin{bmatrix} 1 \\ 2 \\ 0 \\ 0 \end{bmatrix}, \begin{bmatrix} 1 \\ 0 \\ 0 \\ 2 \end{bmatrix}, \begin{bmatrix} 0 \\ 0 \\ 1 \\ 0 \end{bmatrix} \right\}$

   c) Yes a linear transformation, $Ker(T) = span\left\{ \begin{bmatrix} 1 \\ 1 \\ 0 \\ 1 \end{bmatrix}, \begin{bmatrix} 1 \\ 1 \\ 1 \\ 0 \end{bmatrix} \right\}$

   d) Not a linear transformation 　　　　　 e) Not a linear transformation

18 a) Yes, a linear trans. 　　　　　 b) Not a linear trans. $T(0,0) \neq (0,0,0)$

   c) Yes, a lin. trans. 　　　　 d) Yes, a lin. trans.

   e) Not a lin. trans. $T(1,2) + T(2,3) \neq T(3,5)$

19 $Ker(T) = \{(0,0)\}$

20 a) $\begin{bmatrix} 1 & 0 & -1 \\ -4 & 2 & 0 \\ 0 & 1 & 1 \\ 1 & -1 & 0 \\ 0 & 5 & 0 \end{bmatrix}$ b) $\begin{bmatrix} 2 & -3 & -4 \\ -1 & -1 & -5 \\ 0 & 4 & 0 \\ 3 & 0 & 1 \end{bmatrix}$

21 a Yes it forms a basis, since the matrix formed by these vectors is invertible

   b Does not form a basis, since the matrix formed by these vectors is not invertible

   c Does not form a basis, since only two pivot entries

   d Does not form a basis, since the vectors are dependent

22 a)$\{v_1, v_2, v_4\}$ b)$\{v_1, v_2, v_3, v_5\}$ c)$\{v_1, v_2, v_3\}$

23 a) $\{v_1, v_2\}$ b) $\{v_1, v_2, v_3, v_4\}$

24 Null space $= span\left\{ \begin{bmatrix} 5 \\ 4 \\ 1 \\ 0 \\ 0 \end{bmatrix}, \begin{bmatrix} -7 \\ -6 \\ 0 \\ 3 \\ 1 \end{bmatrix} \right\}$

25 a) Columns C_1 and C_2 b) Columns C_1, C_3, C_5

26 a) false b) True c) False

27 $\begin{bmatrix} 1 \\ 5 \end{bmatrix}$

28 Basis $= \{v_2, v_3\}$

29 independent since Corresponding coordinate vectors $\{(1,0,1),(1,0,-1)\}$ is indep.

30 Matrix formed by these vectors is invertible

31 matrix formed by these vectors is not invertible

32 $\{(1,2,3,0),(0,1,4/3,4/3)$

33 C_1, C_2

34 $\begin{bmatrix} -7 \\ 1 \\ 2 \end{bmatrix}$

35 a) $\begin{bmatrix} 3 \\ -7 \end{bmatrix}$ b) $\begin{bmatrix} 2 \\ 5 \end{bmatrix}$ c) $\begin{bmatrix} -1 \\ -5 \\ 9 \end{bmatrix}$ d) $\begin{bmatrix} 0 \\ 1 \\ -5 \end{bmatrix}$

36 a) $\begin{bmatrix} -6 \\ 2 \end{bmatrix}$ b) $\begin{bmatrix} -1 \\ -1 \\ 3 \end{bmatrix}$ c) $\begin{bmatrix} -2 \\ 0 \\ 5 \end{bmatrix}$

37 $\begin{bmatrix} 7 \\ -3 \\ -2 \end{bmatrix}$

38 a) $\begin{bmatrix} 2 & 1 \\ -9 & 8 \end{bmatrix}$ b) $\begin{bmatrix} 3 & 2 & 8 \\ -1 & 0 & -2 \\ 4 & 5 & 7 \end{bmatrix}$

39 $[x]_B = \begin{bmatrix} -7 \\ 5 \end{bmatrix}$

40 $\begin{bmatrix} 5 \\ 1 \\ -2 \end{bmatrix}$

41 The corresponding vectors $\{(1,0,1),(1,0,-3),(1,1,-3)\}$ are independent.

43 a) 2,2 b) 2,3 c) 3,3 d) 0,3 e) 1,2

45 a) false b) True c) True d) False e) False f) False

46　a) $rank = 2$, dim Nul A $= 2$, Basis for Col A $= \left\{ \begin{bmatrix} 1 \\ -1 \\ 5 \end{bmatrix}, \begin{bmatrix} -4 \\ 2 \\ -6 \end{bmatrix} \right\}$ Basis for row space

$= \{(1, -4, 9, -7), (0, -2, 5, -6)\}$ Basis for Nul A $= \left\{ \begin{bmatrix} 1 \\ \frac{5}{2} \\ 1 \\ 0 \end{bmatrix}, \begin{bmatrix} -5 \\ -3 \\ 0 \\ 1 \end{bmatrix} \right\}$

b) $rank = 3$, dim Nul A $= 2$, Basis for Col A $= \left\{ \begin{bmatrix} 1 \\ -2 \\ -3 \\ 3 \end{bmatrix}, \begin{bmatrix} 4 \\ -6 \\ -6 \\ 4 \end{bmatrix}, \begin{bmatrix} 9 \\ -10 \\ -3 \\ 0 \end{bmatrix} \right\}$ Basis for row

space $= \{(1, -3, \frac{4}{3}, 3, 0), (0, 0, 1, \frac{-3}{2}, 3), (0, 0, 0, 0, 1)\}$ Basis for Nul A $= \left\{ \begin{bmatrix} 3 \\ 1 \\ 0 \\ 0 \\ 0 \end{bmatrix}, \begin{bmatrix} -5 \\ 0 \\ \frac{3}{2} \\ 1 \\ 0 \end{bmatrix} \right\}$

c) $rank = 3$, dim Nul A $= 3$, Basis for Col A $= \left\{ \begin{bmatrix} 1 \\ 1 \\ 1 \\ 1 \\ 1 \end{bmatrix}, \begin{bmatrix} 1 \\ 2 \\ -1 \\ -3 \\ -2 \end{bmatrix}, \begin{bmatrix} 7 \\ 10 \\ 1 \\ -5 \\ 0 \end{bmatrix} \right\}$

Basis for row space $= \{(1, -2, 0, 0, -5, -4), (0, 1, -1, \frac{5}{2}, \frac{9}{2}, -2), (0, 0, 0, 1, -1, -2)\}$

Basis for Nul A $= \left\{ \begin{bmatrix} 2 \\ 1 \\ 1 \\ 0 \\ 0 \\ 0 \end{bmatrix}, \begin{bmatrix} -14 \\ -7 \\ 0 \\ 1 \\ 1 \\ 0 \end{bmatrix}, \begin{bmatrix} -2 \\ -3 \\ 0 \\ 2 \\ 0 \\ 1 \end{bmatrix} \right\}$

d) $rank = 2$, dim Nul A $= 3$, Basis for Col A $= \left\{ \begin{bmatrix} 2 \\ 1 \\ -7 \\ 4 \end{bmatrix}, \begin{bmatrix} -1 \\ -2 \\ 8 \\ -5 \end{bmatrix} \right\}$

Basis for row space $= \{(1, \frac{-5}{4}, \frac{-7}{4}, 0, 1), (0, 1, 3, -4, 4)\}$

Basis for Nul A $= \left\{ \begin{bmatrix} -2 \\ -3 \\ 1 \\ 0 \\ 0 \end{bmatrix}, \begin{bmatrix} 5 \\ 4 \\ 0 \\ 1 \\ 0 \end{bmatrix}, \begin{bmatrix} -6 \\ -4 \\ 0 \\ 0 \\ 1 \end{bmatrix} \right\}$

47　a) dim Nul A $= 5$, dim Row space of A$= 3$, rank of $A^T = 3$

b) dim Nul A $= 0$, dim Row space of A$= 3$, rank of $A^T = 3$

c) Col A$= \mathbb{R}^4$, Nul A $\neq \mathbb{R}^3$ since it is a subspace of $\mathbb{R}^7$　　　　　　d) dim Col A$=2$

Chapter 2

Eigenvalues and Eigenvectors

2.1   Introduction

We have observed that a linear transformation $x \longmapsto Ax$ transforms a vector x into another vector Ax. The goal of this chapter is to study this action of a linear transformation. In particular, we study equations such as $Ax = x$, $Ax = 2x$, $Ax = -4x$, $Ax = \frac{1}{2}x$ etc. In general, we study the matrix equations $Ax = \lambda x$ for some scalar λ.

The basic concepts described in this chapter are useful throughout pure and applied mathematics. In particular, they are used in discrete dynamical systems, continuous dynamical system, Engineering design and applications of Physics and Chemistry in a broader sense.

2.2   Eigenvalues and Eigenvectors:

The word **"eigen"** is a German word. It means **"Proper"**, **"Inherent"**, **"Own"**, **"Special"**, **"Peculiar"**, or **"Characteristic"**,

In this section we wish to find proper, characteristic values denoted by **"λ"** such that $Ax = \lambda x$ for some scalar λ.

Definition 2.1. Eigenvector and Eigenvalue: *An **eigenvector** of a $n \times n$ square matrix A is a non-zero vector* $\mathbf{x}$ *such that* $\mathbf{Ax} = \lambda \mathbf{x}$ *for some scalar λ. A scalar λ is called an **eigenvalue** of A if there is a nontrivial solution x of* $\mathbf{Ax} = \lambda \mathbf{x}$; *such a vector* $\mathbf{x}$ *is called an eigenvector corresponding to λ.*

The vector equation $\mathbf{Ax} = \lambda \mathbf{x}$ is very important to determine if a given vector $\mathbf{x}$ is an eigenvector of a square matrix A or not.

Also for a specific value of a scalar λ, it is easy to decide whether it is an eigenvalue or not.

Example 2.1. *Let* $A = \begin{bmatrix} 1 & 6 \\ 5 & 2 \end{bmatrix}$, *for given $\lambda = 7$, we show that λ is an eigenvalue of A as follows.*

Solution: λ will be an eigenvalue of A if the equation $Ax = \lambda x$ has a non-trivial solution. Consider,

$$Ax \qquad\qquad , \; Ax \qquad\qquad Ax \qquad\qquad (A \qquad)$$

We solve the above homogeneous linear system as follows:

$$A - 7I = \begin{bmatrix} 1 & 6 \\ 5 & 2 \end{bmatrix} - 7 \begin{bmatrix} 1 & 0 \\ 0 & 1 \end{bmatrix} = \begin{bmatrix} -6 & 6 \\ 5 & -5 \end{bmatrix}$$

The augmented matrix is

$$\begin{bmatrix} -6 & 6 & 0 \\ 5 & -5 & 0 \end{bmatrix}.$$

Reducing the augmented matrix we get

$$\begin{bmatrix} 1 & -1 & 0 \\ 0 & 0 & 0 \end{bmatrix}.$$

Hence $x_1 = x_2$ is the only linear equation corresponding to the above echelon form. Hence the general solution is $\{t, t\}, t \in \mathbb{R}$.

This shows that the homogeneous system $(A - \lambda I)x = 0$ has a non-trivial solution. This concludes that $\lambda = 7$ is an eigenvalue of A and the corresponding eigen vector is $\begin{bmatrix} 1 \\ 1 \end{bmatrix}$.

This also implies that the vector $\begin{bmatrix} 1 \\ 1 \end{bmatrix}$ forms a basis of the eigenspace corresponding to the eigen value $\lambda = 7$.

Geometrically, this shows that the eigenspace formed for $\lambda = 7$ consists of the vectors which are multiple of $\begin{bmatrix} 1 \\ 1 \end{bmatrix}$.

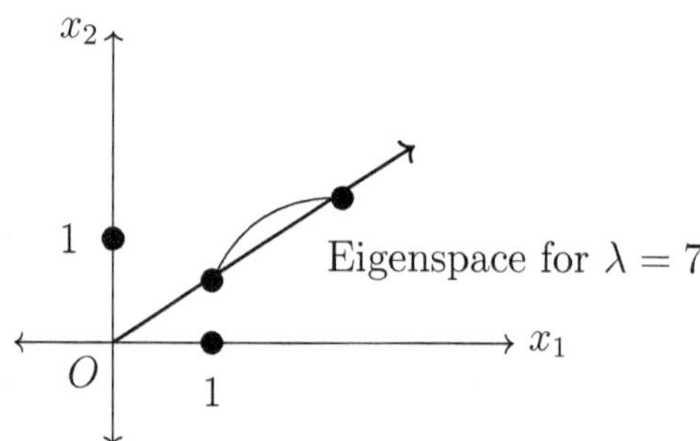

Example 2.2. *Let $A = \begin{bmatrix} 3 & 0 & -1 \\ 2 & 3 & 1 \\ -3 & 4 & 5 \end{bmatrix}$, for given $\lambda = 4$, we test whether λ is an eigenvalue of A or not. If so then we find the corresponding eigenvector.*

Solution: We know that λ will be an eigenvalue of A iff the equation $\mathbf{Ax} = \lambda \mathbf{x}$ has a non-trivial solution.

i.e. $\star Ax = 4x$ has a non-trivial solution. Substitute values in above equation.

$$\left(\begin{bmatrix} 3 & 0 & -1 \\ 2 & 3 & 1 \\ -3 & 4 & 5 \end{bmatrix} - 4 \begin{bmatrix} 1 & 0 & 0 \\ 0 & 1 & 0 \\ 0 & 0 & 1 \end{bmatrix} \right) x = 0$$

Row reduction of the augmented matrix gives

$$\begin{bmatrix} -1 & 0 & -1 & 0 \\ 2 & -1 & 1 & 0 \end{bmatrix} \sim \begin{bmatrix} 1 & 0 & 1 & 0 \\ 0 & 1 & 1 & 0 \end{bmatrix}$$

Rewriting the equations we get

$$x_1 + x_3 = 0$$

$$x_2 + x_3 = 0$$

$$x_3 \quad \text{is free}$$

$$\text{Hence} \quad \begin{bmatrix} x_1 \\ x_2 \\ x_3 \end{bmatrix} = \begin{bmatrix} -x_3 \\ -x_3 \\ x_3 \end{bmatrix} = x_3 \begin{bmatrix} -1 \\ -1 \\ 1 \end{bmatrix}$$

This shows that $\lambda = 4$ is an eigen value of matrix A and the corresponding eigen vector is $\begin{bmatrix} -1 \\ -1 \\ 1 \end{bmatrix}$.

The eigen space is a one - dimensional subspace of $\mathbb{R}^3$ with a basis vector $\begin{bmatrix} -1 \\ -1 \\ 1 \end{bmatrix}$.

In the above examples and discussion we have described the method of finding eigen vectors whenever an eigen value is given.

The following theorem describes a special case in which eigen values are found by inspection for a special type of a matrix A.

Theorem 2.1. *The eigen values of a triangular matrix are the entries on its main diagonal.*

That is if $A = \begin{bmatrix} a_{11} & a_{12} & a_{13} \\ 0 & a_{22} & a_{23} \\ 0 & 0 & a_{33} \end{bmatrix}$ then the eigen values are a_{11}, a_{22} and a_{33}.

Example 2.3. *Given matrices A are*

$$\text{i)} \quad A = \begin{bmatrix} 3 & 6 & -8 \\ 0 & 0 & 6 \\ 0 & 0 & 2 \end{bmatrix} \qquad \text{ii)} \quad A = \begin{bmatrix} 4 & 0 & 0 \\ -2 & 1 & 0 \\ 5 & 3 & 4 \end{bmatrix}$$

Solution:

(i) Matrix A has $3, 0, 2$ as eigen values 0 is an eigen value means the equation $Ax = 0x$ has a non-trivial solution.

That is possible only if matrix A is not invertible. Thus 0 is an eigen value of A iff A is not invertible.

(ii) Matrix **A** has $4, 1, 4$ as eigen values. The eigen value 4 is repeated.

The following theorem states about the linear independence of the set of eigen vectors.

Theorem 2.2. *If $v_1, v_2, \cdots, v_r$ are eigen vectors that correspond to distinct eigen values $\lambda_1, \lambda_2 \cdots \lambda_r$ of an $n \times n$ matrix A, then the set $\{v_1, v_2, \cdots, v_r\}$ is linearly independent.*

Example 2.4. *The eigen values of the triangular matrix $A = \begin{bmatrix} 0 & 0 & 0 \\ 0 & 2 & 5 \end{bmatrix}$ are $0\ 2\ -1$*

Solution: The eigenvalues are distinct. It is easy to see that the eigenvectors corresponding to $0, 2, -1$ values are $\begin{bmatrix} 1 \\ 0 \\ 0 \end{bmatrix}$, $\begin{bmatrix} 0 \\ 1 \\ 0 \end{bmatrix}$ and $\begin{bmatrix} 0 \\ 0 \\ 1 \end{bmatrix}$ respectively. So we apply theorem 2.2 to conclude that the set of eigenvectors

$$\left\{ \begin{bmatrix} 1 \\ 0 \\ 0 \end{bmatrix}, \begin{bmatrix} 0 \\ 1 \\ 0 \end{bmatrix}, \begin{bmatrix} 0 \\ 0 \\ 1 \end{bmatrix} \right\}$$

is linearly independent.

Example 2.5. *Let* $A = \begin{bmatrix} 1 & -1 & 0 \\ 2 & 3 & 2 \\ 1 & 1 & 2 \end{bmatrix}$, *Test whether the vector* $x = \begin{bmatrix} -2 \\ 2 \\ 1 \end{bmatrix}$ *is an eigenvector of A, if so, find eigenvalue.*

Solution: Compute $Ax = \begin{bmatrix} -4 \\ 4 \\ 2 \end{bmatrix} = 2 \begin{bmatrix} -2 \\ 2 \\ 1 \end{bmatrix} = 2x$. Hence the given vector x is an eigen vector and the eigen value is 2.

2.3 The Characteristic Equation:

In this section, we discuss the method of finding eigen values of a matrix A. This is done by forming a special scalar equation called **the characteristic equation of** A. We illustrate the method with the help of examples.

Example 2.6. *Find the eigen values of the matrix* $A = \begin{bmatrix} 2 & 3 \\ 3 & -6 \end{bmatrix}$.

Solution: We find scalars λ such that they satisfy the vector equation $(A - \lambda I)x = O$.

$$A - \lambda I = \begin{bmatrix} 2 & 3 \\ 3 & -6 \end{bmatrix} - \begin{bmatrix} \lambda & 0 \\ 0 & \lambda \end{bmatrix} = \begin{bmatrix} 2 - \lambda & 3 \\ 3 & -6 - \lambda \end{bmatrix}$$

$$\therefore (A - \lambda I)x = \begin{bmatrix} 2 - \lambda & 3 \\ 3 & -6 - \lambda \end{bmatrix} x$$

We find λ of matrix $A - \lambda I$ by computing $\det(A - \lambda I) = O$. That is

$$\begin{vmatrix} 2 - \lambda & 3 \\ 3 & -6 - \lambda \end{vmatrix} = (2 - \lambda)(-6 - \lambda) - 9 = 0$$

$$= \lambda^2 + 4\lambda - 21 = 0$$

$$therefore (\lambda - 3)(\lambda + 7) = 0$$

$$\therefore \lambda = 3 \text{ or } \lambda = -7$$

Hence the eigen values are 3 and -7.

For square matrices of order higher than or equal to 3×3 we use the following fact about det **A** to find λ values.

- Let A be a given $n \times n$ matrix.

- Let U be any echelon form of A obtained by row reduction.

- Suppose r is the number of row interchanges used during row reduction.

- Then the determinant A is $(-1)^r$ times the product of diagonals entries $u_{11}, u_{22}, \cdots, u_{nn}$.

- If A is an invertible matrix then $A \sim I_n$ and $\det A = u_{11}\, u_{22} \cdots u_{nn}$ with all pivot elements (not all $= 1$),otherwise at least one of them is zero. Hence the product $u_{11}, u_{22} \cdots u_{nn}$ is zero.

- Thus

$$\det A = \begin{cases} (-1)^r \left(\begin{array}{c} \text{product of} \\ \text{pivot elements in } U \end{array} \right) & \text{when } A \text{ is invertible} \\ 0 & \text{when } A \text{ is } \textbf{not} \text{ invertible} \end{cases}$$

Example 2.7. *Compute* $\det A$ *of the following matrices using steps described above.*

Solution:

(i) $A = \begin{bmatrix} 1 & 5 & 0 \\ 2 & 4 & -1 \\ 0 & -2 & 0 \end{bmatrix}$

First we row reduce matrix A to obtain U.

$$\begin{bmatrix} 1 & 5 & 0 \\ 2 & 4 & -1 \\ 0 & -2 & 0 \end{bmatrix} \sim \begin{bmatrix} 1 & 5 & 0 \\ 0 & -6 & -1 \\ 0 & 0 & \frac{1}{3} \end{bmatrix} = U$$

This reduction uses no row interchange operation, that is $r = 0$.

Therefore, $\det A = (-1)^0 (1)(-6) \left(\frac{1}{3} \right) = -2$.

(ii) $A = \begin{bmatrix} 0 & 3 & 1 \\ 3 & 0 & 2 \\ 1 & 2 & 0 \end{bmatrix}$

$$A\, R_{13} \sim \begin{bmatrix} 1 & 2 & 0 \\ 3 & 0 & 2 \\ 0 & 3 & 1 \end{bmatrix} R_2 - 3R_1 \sim \begin{bmatrix} 1 & 2 & 0 \\ 0 & -6 & 2 \\ 0 & 3 & 1 \end{bmatrix} R_3 + \tfrac{1}{2}R_2 \sim \begin{bmatrix} 1 & 2 & 0 \\ 0 & -6 & 2 \\ 0 & 0 & 2 \end{bmatrix} = U$$

We obtain matrix U with one iterchange. So, $r = 1$.

$\therefore \det A = (-1)^1 (1)(-6)(2) = 12$

(iii) $A = \begin{bmatrix} 3 & -2 \\ 1 & -1 \end{bmatrix}$

$$\begin{bmatrix} 3 & -2 \\ 1 & -1 \end{bmatrix} R_2 - \tfrac{1}{3}R_1 \sim \begin{bmatrix} 3 & -2 \\ 0 & \frac{-1}{3} \end{bmatrix}, \text{ No interchange of rows.}$$

$$\therefore \det A = (-1)^0 (3) \left(\frac{1}{3} \right) = 1.$$

Theorem 2.3. *A square matrix* A *of order* n *is invertible iff*

(b) The determinant of A is **not zero.**

Geometrically, when A is a 3×3 matrix, $\det A$ denotes the volume of the parallelepiped determined by the columns of $A, \mathbf{a_1}, \mathbf{a_2}, \mathbf{a_3}$.

This volume is non-zero iff the vectors a_1, a_2, a_3 **are linearly independent.** In this case matrix A is invertible.

If a_1, a_2, a_3 are non-zero and linearly dependent then they lie in a plane or along a line.

Definition 2.2. Characteristic Equation of A

Let A be an $n \times n$ matrix. The scalar equation $\det(A - \lambda I) = 0$ *is called* **the characteristic equation** **of** A.

Note that, λ is an eigenvalue of a square matrix A iff λ satisfies the characteristic equation

$$\det(A - \lambda I) = 0.$$

The equation $\det(A - \lambda I) = 0$ is a polynomial in λ.

We factorize the polynomial to obtain the eigenvalues corresponding to the given matrix A.

Example 2.8. *Find the characteristic polynomial and eigenvalues of the following matrices.*

$(a)\ A = \begin{bmatrix} 2 & 7 \\ 7 & 2 \end{bmatrix}$
$(b)\ A = \begin{bmatrix} 4 & 0 & 0 \\ 5 & 3 & 2 \\ -2 & 0 & 2 \end{bmatrix}$
$(c)\ A = \begin{bmatrix} 4 & -7 & 0 & 2 \\ 0 & 3 & -4 & 6 \\ 0 & 0 & 3 & -8 \\ 0 & 0 & 0 & 1 \end{bmatrix}$

Solution: (a)

$$\det(A - \lambda I) = (2 - \lambda)(2 - \lambda) - 49$$
$$= (2 - \lambda)^2 - 49$$
$$= ((2 - \lambda) - 7)(2 - \lambda + 7)$$
$$= (-5 - \lambda)(9 - \lambda)$$

$$\det(A - \lambda I) = 0 \textbf{ gives } \lambda = -5 \text{ and } \lambda = 9.$$

(b)

$$\det(A - \lambda I) = (4 - \lambda) \begin{vmatrix} 3 - \lambda & 2 \\ 0 & 2 - \lambda \end{vmatrix}$$
$$= (4 - \lambda)(3 - \lambda)(2 - \lambda) = 0$$

So the eigenvalues are $4, 3, 2$.

(c) Matrix A is triangular, so is $A - \lambda I$.

$$\therefore \det(A - \lambda I) = \text{product of diagonal elements}$$
$$= (4 - \lambda)(3 - \lambda)(3 - \lambda)(1 - \lambda) = 0$$

2.4 Diagonalization

Definition 2.3. Similar Matrices

Let $\mathbf{A}$ and $\mathbf{B}$ be square matrices of order n. Matrix $\mathbf{A}$ is said to be similar to matrix $\mathbf{B}$ if there is an invertible matrix $\mathbf{P}$ such that $\mathbf{P}^{-1}\mathbf{A}\mathbf{P} = \mathbf{B}$. (Equivalently $A = PBP^{-1}$)

Example 2.9. *The matrices $A = \begin{bmatrix} 7 & 2 \\ -4 & 1 \end{bmatrix}$ and $B = \begin{bmatrix} 5 & 0 \\ 0 & 3 \end{bmatrix}$ are similar.*

Solution: Let $P = \begin{bmatrix} 1 & 1 \\ -1 & -2 \end{bmatrix}$. Then $P^{-1} = \begin{bmatrix} 2 & 1 \\ -1 & -1 \end{bmatrix}$.
It is easy to see that $P^{-1}AP = B$. Hence matrices A and B are said to be similar.

Theorem 2.4. *If $n \times n$ matrices A and B are similar, then they have the same characteristic polynomial and hence the same eigenvalues (with same multiplicity).*

Note that the converse may not be true. For example, $A = \begin{bmatrix} 2 & 1 \\ 0 & 2 \end{bmatrix}$, $B = \begin{bmatrix} 2 & 0 \\ 0 & 2 \end{bmatrix}$.
Here $A \sim B, A$ and B have same eigen values but A is not similar to B. (verify).

Definition 2.4. A diagonalizable matrix

A square matrix A is said to be diagonalizable if A is similar to a diagonal matrix, that is $A = PDP^{-1}$ for some invertible matrix P and some diagonal matrix D.

The next theorem is a characterization of diagonalizable matrices.

Theorem 2.5. (The Diagonalization Theorem)

An $n \times n$ matrix A is diagonalizable if and only if A has n linearly independent eigenvectors. That is, $A = PDP^{-1}$, with D a diagonal matrix if and only if the columns of P are n linearly independent eigen vectors of A.

That is, matrix A is diagonalizable iff there are enough eigen vectors to form a basis of $\mathbb{R}^n$.
The following example illustrates the process of diagonalization of a given matrix A.

Example 2.10. *Let $A = \begin{bmatrix} 1 & -3 & 3 \\ 3 & -5 & 3 \\ 6 & -6 & 4 \end{bmatrix}$*

Solution: For the given matrix A we first find the eigen values by forming a characteristic equation

$$|A - \lambda I| = \begin{vmatrix} 1 - \lambda & -3 & 3 \\ 3 & -5 - \lambda & 3 \\ 6 & -6 & 4 - \lambda \end{vmatrix} = \lambda^3 - 8\lambda^2 - 12\lambda - 16$$

$$= (\lambda + 2)(\lambda + 2)(\lambda - 4)$$

$$= 0$$

In the second step we find eigen vectors coresponding to distinct eigen values.

For $\lambda = -2$

$$A - \lambda I = \begin{bmatrix} 1+2 & -3 & 3 \\ 3 & -5+2 & 3 \\ 6 & -6 & 6 \end{bmatrix} = \begin{bmatrix} +3 & -3 & 3 \\ 3 & -3 & 3 \\ 6 & -6 & 6 \end{bmatrix}$$

Now, row reduce the augmented mateix for $(A - (-2)I)x = 0$.

$$\begin{bmatrix} +3 & -3 & 3 & 0 \\ 3 & -3 & 3 & 0 \\ 6 & -6 & 6 & 0 \end{bmatrix} \sim \begin{bmatrix} 3 & -3 & 3 & 0 \\ 0 & 0 & 0 & 0 \\ 0 & 0 & 0 & 0 \end{bmatrix} \sim \begin{bmatrix} 1 & -1 & 1 & 0 \\ 0 & 0 & 0 & 0 \\ 0 & 0 & 0 & 0 \end{bmatrix}$$

The corresponding eigen vectors are computed as follows:

$$\begin{bmatrix} x_1 \\ x_2 \\ x_3 \end{bmatrix} = \begin{bmatrix} x_2 - x_3 \\ x_2 \\ x_3 \end{bmatrix} = x_2 \underset{\underset{P_1}{\uparrow}}{\begin{bmatrix} 1 \\ 1 \\ 0 \end{bmatrix}} + x_3 \underset{\underset{P_2}{\uparrow}}{\begin{bmatrix} -1 \\ 0 \\ 1 \end{bmatrix}}$$

For $\lambda = 4$

$$A - \lambda I = \begin{bmatrix} 1-4 & -3 & 3 \\ 3 & -5-4 & 3 \\ 6 & -6 & 4-4 \end{bmatrix} = \begin{bmatrix} -3 & -3 & 3 \\ 3 & -9 & 3 \\ 6 & -6 & 0 \end{bmatrix}$$

Now, row reduce the augmented matrix for $(A - 4I)\,x = 0$

$$\begin{bmatrix} -3 & -3 & 3 & 0 \\ 3 & -9 & 3 & 0 \\ 6 & -6 & 0 & 0 \end{bmatrix} \sim \begin{bmatrix} 1 & 1 & -1 & 0 \\ 0 & -2 & 1 & 0 \\ 0 & 0 & 0 & 0 \end{bmatrix}$$

The corresponding eigen vectors are computed as follows:

$$\begin{bmatrix} x_1 \\ x_2 \\ x_3 \end{bmatrix} = \begin{bmatrix} \frac{1}{2}x_3 \\ \frac{1}{2}x_3 \\ x_3 \end{bmatrix} = x_3 \underset{\underset{P_3}{\uparrow}}{\begin{bmatrix} \frac{1}{2} \\ \frac{1}{2} \\ 1 \end{bmatrix}}$$

It is easy to see that P_1, P_2, P_3 are linearly independent.

We have

$$P = [P_1 \, P_2 \, P_3] = \begin{bmatrix} 1 & -1 & \frac{1}{2} \\ 1 & 0 & \frac{1}{2} \\ 0 & 1 & 1 \end{bmatrix}$$

Now as a last step we show that matrices A and $D = \begin{bmatrix} -2 & 0 & 0 \\ 0 & -2 & 0 \\ 0 & 0 & 4 \end{bmatrix}$ are similar matrices.

That is we show $AP = PD$. This avoids calculating P^{-1}.

$$AP = \begin{bmatrix} -2 & 2 & 2 \\ -2 & 0 & 2 \end{bmatrix} = PD$$

$$\text{Hence } P^{-1}AP = D \text{ or } A = PDP^{-1}.$$

In this way we have diagonalized the given matrix.

In the above example note that one eigen value is repeated. In other words there are only two distinct eigenvalues. However the number of distinct eigenvectors is three.

In some cases it may happen that if an eigen value is repeated, the number of eigen vectors is less than the order of the matrix A. In such cases the matrix P can not exist, as a result A will not be diagonalizable.

The following theorem states the assurance of diagonalizability of matrix A whenever matrix A has distinct eigen values.

Theorem 2.6. *An $n \times n$ matrix with n distinct eigen values is diagonalizable.*

Example 2.11. *Diagonalize the matrix A if possible, where $A = \begin{bmatrix} -1 & 4 & 2 \\ -3 & 4 & 0 \\ -3 & 1 & 3 \end{bmatrix}$.*

Solution: Note that, characteristic equation has three distinct roots, as $\lambda = 1, 2, 3$. So from above theorem, we are sure that A is diagonalizable.

We get $P_1 = \begin{bmatrix} 1 \\ 1 \\ 1 \end{bmatrix}$, $P_2 = \begin{bmatrix} \frac{2}{3} \\ 1 \\ 1 \end{bmatrix}$, $P_3 = \begin{bmatrix} \frac{1}{4} \\ \frac{3}{4} \\ 1 \end{bmatrix}$

Now it remains to show that $AP = PD$, where $P = \begin{bmatrix} 1 & \frac{2}{3} & \frac{1}{4} \\ 1 & 1 & \frac{3}{4} \\ 1 & 1 & 1 \end{bmatrix}$ and $D = \begin{bmatrix} 1 & 0 & 0 \\ 0 & 2 & 0 \\ 0 & 0 & 3 \end{bmatrix}$.

Simple calculation shows that

$$AP = PD = \begin{bmatrix} 1 & 1\cdot 33 & 0\cdot 75 \\ 1 & 2 & 2\cdot 25 \\ 1 & 2 & 3 \end{bmatrix}.$$

2.5 Eigenvectors and linear transformation

The square matrix A of order n, written as $A = PDP^{-1}$ is nothing but matrix factorization of A. The transformation $x \longmapsto Ax$ is essentially the same as the very simple mapping $x \longmapsto Dx$ when viewed with a proper perspective. We know that any linear transformation T from $\mathbb{R}^n$ to $\mathbb{R}^m$ is a matrix transformation. If A is diagonalizable, then there is a basis $\mathbb{B}$ for $\mathbb{R}^n$ consisting of eigen vectors of A. In this case the $\mathbb{B}$-matrix of T is diagonal. Hence diagonalizing A amounts to finding a diagonal matrix representation of the map $x \longmapsto Ax$. The following theorem tells about the $\mathbb{B}$-matrix for the matrix transformation $x \longmapsto Ax$.

Theorem 2.7. (Diagonal matrix representation)

columns of P then D is the $\mathbb{B}$-matrix for the transformation $x \longmapsto Ax$.

Proof: Let the columns of matrix $\mathbf{P}$ be $\mathbf{b_1}, \mathbf{b_2}, \cdots, \mathbf{b_n}$. That is the basis $\mathbb{B}$ for $\mathbb{R}^n$ is $\{\mathbf{b_1}, \mathbf{b_2}, \cdots, \mathbf{b_n}\}$ and $\mathbf{P} = [\mathbf{b_1}, \mathbf{b_2}, \cdots, \mathbf{b_n}]$.

This matrix $\mathbf{P}$ is responsible for the corresponding coordinate mapping $x \longmapsto [x]_{\mathbb{B}}$ discussed in first chapter. That is, $\mathbf{P}[\mathbf{x}]_{\mathbb{B}} = \mathbf{x}$ and $[\mathbf{x}]_{\mathbb{B}} = \mathbf{P}^{-1}\mathbf{x}$

Let $T : \mathbb{R}^n \to \mathbb{R}^n$ be a linear transformation defined as $x \longmapsto Ax$. Then the image of basis $\mathbb{B}$ under T is

$$\begin{aligned}
[T]_{\mathbb{B}} &= [[(T(b_1)]_{\mathbb{B}}\,[T(b_2)]_{\mathbb{B}} \cdots [T(b_n)]_{\mathbb{B}}] \\
&= [[Ab_1]_{\mathbb{B}}\,[Ab_2]_{\mathbb{B}} \cdots [Ab_n]_{\mathbb{B}}] \\
&= [P^{-1}Ab_1 \; P^{-1}Ab_2 \cdots P^{-1}Ab_n] \qquad \text{(Coordinate mapping)} \\
&= P^{-1}A[b_1\,b_2\cdots b_n] \qquad\qquad\quad \text{(matrix multiplication)} \\
&= P^{-1}AP = D
\end{aligned}$$

$$\text{Hence } [T]_{\mathbb{B}} = D$$

This shows that $\mathbf{D}$ is the $\mathbb{B}$-matrix for the linear transformation $x \longmapsto Ax$. The following examples use theorem 2.7.

Example 2.12. *Let $T : \mathbb{R}^2 \to \mathbb{R}^2$ be $x \longmapsto Ax$ transformation where $A = \begin{bmatrix} 7 & 2 \\ -4 & 1 \end{bmatrix}$. To find basis $\mathbb{B}$ of $\mathbb{R}^2$ such that the $\mathbb{B}$ matrix for T is a diagonal matrix.*

Solution: As a first step of the solution to this problem we check whether A is diagonalizable. It is easy to see that A is diagonalizable where $\mathbf{P} = \begin{bmatrix} 1 & 1 \\ -1 & -2 \end{bmatrix}$ and $\mathbf{D} = \begin{bmatrix} 5 & 0 \\ 0 & 3 \end{bmatrix}$.

Let $\mathbf{b_1} = \begin{bmatrix} 1 \\ -1 \end{bmatrix}$ and $\mathbf{b_2} = \begin{bmatrix} 1 \\ -2 \end{bmatrix}$.

Therefore, $\mathbb{B} = \{b_1, b_2\}$ is the required basis of $\mathbb{R}^2$ such that T is defined in terms of a diagonal matrix D **by Diagonal matrix Representation theorem.**

Example 2.13. *Find the $\mathbb{B}$-matrix for the transformation $x \longmapsto Ax$ when $\mathbb{B} = \{b_1, b_2\}$ for given* $A = \begin{bmatrix} 3 & 4 \\ -1 & -1 \end{bmatrix}$, $b_1 = \begin{bmatrix} 2 \\ -1 \end{bmatrix}$, $b_2 = \begin{bmatrix} 1 \\ 2 \end{bmatrix}$.

Solution: Given $P = [b_1, b_2]$. That is $P = \begin{bmatrix} 2 & 1 \\ -1 & 2 \end{bmatrix}$.

So, the required $\mathbb{B}$-matrix is $\mathbf{P}^{-1}\mathbf{AP}$.

Using adjoint definition of inverse of a matrix we get,

$$\mathbf{P}^{-1} = \frac{1}{5}\begin{bmatrix} 2 & -1 \\ 1 & 2 \end{bmatrix} \left\{ P^{-1} = \frac{1}{|P|}\,adj\,p \right\}$$

$$\therefore \mathbf{P}^{-1}\mathbf{AP} = \frac{1}{5}\begin{bmatrix} 2 & -1 \\ 1 & 2 \end{bmatrix}\begin{bmatrix} 3 & 4 \\ -1 & -1 \end{bmatrix}\begin{bmatrix} 2 & 1 \\ -1 & 2 \end{bmatrix}$$

$$= \begin{bmatrix} 1 & 5 \end{bmatrix}$$

Which is a   Required $\mathbb{B}-$matrix

EXERCISE

1. In each of the following problems test whether the given λ value is eigenvalue of given matrix A. If so, find the corresponding eigen vector(s).

   (a) $\lambda = 2$, $A = \begin{bmatrix} 3 & 2 \\ 3 & 8 \end{bmatrix}$

   (b) $\lambda = -2$, $A = \begin{bmatrix} 7 & 3 \\ 3 & -1 \end{bmatrix}$

   (c) $\lambda = 4$, $A = \begin{bmatrix} 3 & 0 & -1 \\ 2 & 3 & 1 \\ -3 & 4 & 5 \end{bmatrix}$

   (d) $\lambda = 3$, $A = \begin{bmatrix} 1 & 2 & 2 \\ 3 & -2 & 1 \\ 0 & 1 & 1 \end{bmatrix}$

   (e) $\lambda = 5$, $A = \begin{bmatrix} 6 & -3 & 1 \\ 3 & 0 & 5 \\ 2 & 2 & 6 \end{bmatrix}$

   (f) $\lambda = -3$, $A = \begin{bmatrix} 4 & 0 & 0 \\ 0 & 0 & 0 \\ 1 & 0 & -3 \end{bmatrix}$

2. Test whether the given vector v is an eigen vector of the given matrix A.

   (a) $v = \begin{bmatrix} 1 \\ 4 \end{bmatrix}$, $A = \begin{bmatrix} -3 & 1 \\ -3 & 8 \end{bmatrix}$

   (b) $v = \begin{bmatrix} -1+\sqrt{2} \\ 1 \end{bmatrix}$, $A = \begin{bmatrix} 2 & 1 \\ 1 & 4 \end{bmatrix}$

   (c) $v = \begin{bmatrix} 4 \\ -3 \\ 1 \end{bmatrix}$, $A = \begin{bmatrix} -3 & 7 & 9 \\ -4 & -5 & 1 \\ 2 & 4 & 4 \end{bmatrix}$

   (d) $v = \begin{bmatrix} 1 \\ -2 \\ 1 \end{bmatrix}$, $A = \begin{bmatrix} 3 & 6 & 7 \\ 3 & 3 & 7 \\ 2 & 6 & 5 \end{bmatrix}$

3. Find a basis for the eigen space corresponding to each of the following:

   (a) $A = \begin{bmatrix} 5 & 0 \\ 2 & 1 \end{bmatrix}$, $\lambda = 1, 5$

   (b) $A = \begin{bmatrix} 10 & -9 \\ 4 & -2 \end{bmatrix}$, $\lambda = 4$

   (c) $A = \begin{bmatrix} 7 & 4 \\ -3 & -1 \end{bmatrix}$, $\lambda = 1, 5$

   (d) $A = \begin{bmatrix} 4 & 0 & 1 \\ -2 & 1 & 0 \\ -2 & 0 & 1 \end{bmatrix}$, $\lambda = 1, 2, 3$

   (e) $A = \begin{bmatrix} 1 & 0 & -1 \\ 1 & -3 & 0 \\ 4 & -13 & 1 \end{bmatrix}$, $\lambda = -2$

   (f) $A = \begin{bmatrix} 3 & 0 & 2 & 0 \\ 1 & 3 & 1 & 0 \\ 0 & 1 & 1 & 0 \\ 0 & 0 & 0 & 4 \end{bmatrix}$, $\lambda = 4$

4. Find the eigen values of the following matrices.

   (a) $\begin{bmatrix} 5 & -6 & -6 \\ -1 & 4 & 2 \\ 3 & -6 & -4 \end{bmatrix}$

   (b) $\begin{bmatrix} 4 & 0 & 0 \\ 0 & 0 & 0 \\ 1 & 0 & -3 \end{bmatrix}$

   (c) $\begin{bmatrix} 1 & 2 \\ 3 & 2 \end{bmatrix}$

   (d) $\begin{bmatrix} 1 & 2 & 3 \\ 2 & 1 & 3 \\ 3 & 3 & 6 \end{bmatrix}$

   (e) $\begin{bmatrix} 3 & 2 & 2 \\ 1 & 4 & 1 \\ -2 & -4 & -1 \end{bmatrix}$

   (f) $\begin{bmatrix} 7 & 4 & -4 \\ 4 & -8 & -1 \\ -4 & -1 & -8 \end{bmatrix}$

5. Without calculation find one eigen value of $A = \begin{bmatrix} 1 & 2 & 3 \\ 1 & 2 & 3 \\ 1 & 2 & 3 \end{bmatrix}$

6. Find characteristic polynomial and the eigen values of the matrices given below:

(a) $\begin{bmatrix} 5 & 3 \\ -4 & 4 \end{bmatrix}$ (b) $\begin{bmatrix} 7 & -2 \\ 2 & 3 \end{bmatrix}$ (c) $\begin{bmatrix} 2 & 3 & -2 \\ 0 & 5 & 4 \\ 1 & 0 & -1 \end{bmatrix}$

7. Diagonalize the following matrices if possible:

(a) $\begin{bmatrix} 2 & 2 & 1 \\ 1 & 3 & 1 \\ 1 & 2 & 2 \end{bmatrix}$ (b) $\begin{bmatrix} 4 & 0 & -2 \\ 2 & 5 & 4 \\ 0 & 0 & 5 \end{bmatrix}$ (c) $\begin{bmatrix} 1 & 0 \\ 6 & -1 \end{bmatrix}$ (d) $\begin{bmatrix} 5 & 1 \\ 0 & 5 \end{bmatrix}$

(e) $\begin{bmatrix} 3 & -1 \\ 1 & 5 \end{bmatrix}$ (f) $\begin{bmatrix} -1 & 4 & -2 \\ -3 & 4 & 0 \\ -3 & 1 & 3 \end{bmatrix}$ (g) $\begin{bmatrix} 0 & -4 & -6 \\ -1 & 0 & -3 \\ 1 & 2 & 5 \end{bmatrix}$

(h) $\begin{bmatrix} -7 & -16 & 4 \\ 6 & 13 & -2 \\ 12 & 16 & 1 \end{bmatrix}$ (i) $\begin{bmatrix} 5 & -3 & 0 & 9 \\ 0 & 3 & 1 & -2 \\ 0 & 0 & 2 & 0 \\ 0 & 0 & 0 & 2 \end{bmatrix}$ (j) $\begin{bmatrix} 4 & 0 & 0 & 0 \\ 0 & 4 & 0 & 0 \\ 0 & 0 & 2 & 0 \\ 1 & 0 & 0 & 2 \end{bmatrix}$

8. Find the $\mathbb{B}$-matrix for the transformation $x \longmapsto Ax$ when $\mathbb{B} = \{b_1, b_2\}$.

(a) $A = \begin{bmatrix} -1 & 4 \\ -2 & 3 \end{bmatrix}$, $b_1 = \begin{bmatrix} 3 \\ 2 \end{bmatrix}$, $b_2 = \begin{bmatrix} -1 \\ 1 \end{bmatrix}$

(b) $A = \begin{bmatrix} 1 & 1 \\ -1 & 3 \end{bmatrix}$, $b_1 = \begin{bmatrix} 1 \\ 1 \end{bmatrix}$, $b_2 = \begin{bmatrix} 5 \\ 4 \end{bmatrix}$

(c) $A = \begin{bmatrix} -7 & -48 & -16 \\ 1 & 14 & 6 \\ -3 & -45 & -19 \end{bmatrix}$, $b_1 = \begin{bmatrix} -3 \\ 1 \\ -3 \end{bmatrix}$, $b_2 = \begin{bmatrix} -2 \\ 1 \\ -3 \end{bmatrix}$, $b_3 = \begin{bmatrix} 3 \\ -1 \\ 0 \end{bmatrix}$

(d) $A = \begin{bmatrix} -14 & 4 & -14 \\ -33 & 9 & -31 \\ 11 & -4 & 11 \end{bmatrix}$, $b_1 = \begin{bmatrix} -1 \\ -2 \\ 1 \end{bmatrix}$, $b_2 = \begin{bmatrix} -1 \\ -1 \\ 1 \end{bmatrix}$, $b_3 = \begin{bmatrix} -1 \\ -2 \\ 0 \end{bmatrix}$

9. Define $T : \mathbb{R}^2 \to \mathbb{R}^2$ by $T(x) = Ax$. Find a basis $\mathbb{B}$ for $\mathbb{R}^2$ with the property that $[T]_\mathbb{B}$ is diagonal.

(a) $A = \begin{bmatrix} 0 & 1 \\ -3 & 4 \end{bmatrix}$ (b) $A = \begin{bmatrix} 5 & -3 \\ -7 & 1 \end{bmatrix}$ (c) $A = \begin{bmatrix} 4 & -2 \\ -1 & 3 \end{bmatrix}$ (d) $A = \begin{bmatrix} 2 & -6 \\ -1 & 3 \end{bmatrix}$

10. Define $T : \mathbb{R}^4 \to \mathbb{R}^4$ by $T(x) = Ax$. Find a basis $\mathbb{B}$ for $\mathbb{R}^4$ with the property that $[T]_\mathbb{B}$ is diagonal, where

$$A = \begin{bmatrix} 15 & -66 & -44 & -33 \\ 0 & 13 & 21 & -15 \\ 1 & -15 & -21 & 12 \\ 2 & -18 & -22 & 8 \end{bmatrix}$$

SOLUTIONS

1 a) Yes $\begin{bmatrix} -2 \\ 1 \end{bmatrix}$

b) Yes $\begin{bmatrix} -1 \\ 3 \end{bmatrix}$

c) Yes $\begin{bmatrix} -1 \\ -1 \end{bmatrix}$

d) Yes $\begin{bmatrix} 3 \\ 2 \\ 1 \end{bmatrix}$

e) Not an eigenvalue

f) Yes $\begin{bmatrix} 0 \\ 0 \\ 1 \end{bmatrix}$

2 a) Not an eigen vector

b) Yes $\lambda = 3 + \sqrt{(2)}$

c) Yes $\lambda = 0$

d) Not an eigen vector

3 a) $\lambda = 1,\ \begin{bmatrix} 0 \\ 1 \end{bmatrix},\ \lambda = 5,\ \begin{bmatrix} 2 \\ 1 \end{bmatrix}$

b) $\lambda = 4,\ \begin{bmatrix} 3 \\ 2 \end{bmatrix}$

c) $\lambda = 1,\ \begin{bmatrix} -2 \\ 3 \end{bmatrix},\ \lambda = 5,\ \begin{bmatrix} -2 \\ 1 \end{bmatrix}$

d) $\lambda = 1,\ \begin{bmatrix} 0 \\ 1 \\ 0 \end{bmatrix},\ \lambda = 2,\ \begin{bmatrix} 1 \\ -2 \\ -2 \end{bmatrix},\ \lambda = 3,\ \begin{bmatrix} 1 \\ -1 \\ -1 \end{bmatrix}$

e) $\lambda = -2,\ \begin{bmatrix} 1 \\ 1 \\ 3 \end{bmatrix}$

f) $\lambda = 4,\ \begin{bmatrix} 2 \\ 3 \\ 1 \\ 0 \end{bmatrix},\ \begin{bmatrix} 0 \\ 0 \\ 0 \\ 1 \end{bmatrix}$

4 a) 1,2

b) 4,0,-3

c) 4,-1

d) 9,-1,0

e) 1,2,3

f) 9,-9

5 0

6 a) $\lambda^2 - 9\lambda + 32,\ \lambda = \frac{-9+\sqrt{47}i}{2},\ \frac{-9-\sqrt{47}i}{2}$

b) $\lambda^2 - 10\lambda + 25,\ \lambda = 5$

7 a) $P = \begin{bmatrix} 5 & 0 & 0 \\ 0 & 1 & 0 \\ 0 & 0 & 1 \end{bmatrix}, D = \begin{bmatrix} 1 & 1 & 2 \\ 1 & 0 & -1 \\ 1 & -1 & 0 \end{bmatrix}$

b) $P = \begin{bmatrix} 4 & 0 & 0 \\ 0 & 5 & 0 \\ 0 & 0 & 5 \end{bmatrix}, D = \begin{bmatrix} 1 & 1 & 0 \\ -2 & 0 & 1 \\ 0 & \frac{-1}{2} & 0 \end{bmatrix}$

c) $P = \begin{bmatrix} -1 & 0 \\ 0 & 1 \end{bmatrix}, D = \begin{bmatrix} 0 & 1 \\ 1 & 3 \end{bmatrix}$

d) Not diagonalizable

e) Not diagonalizable

f) $P = \begin{bmatrix} 1 & 0 & 0 \\ 0 & 2 & 0 \\ 0 & 0 & 3 \end{bmatrix}, D = \begin{bmatrix} 1 & 2 & 1 \\ 1 & 3 & 3 \\ 1 & 3 & 4 \end{bmatrix}$

g) $P = \begin{bmatrix} 1 & 0 & 0 \\ 0 & 2 & 0 \\ 0 & 0 & 2 \end{bmatrix}, D = \begin{bmatrix} 2 & 3 & 0 \\ 1 & 0 & 3 \\ -1 & -1 & -2 \end{bmatrix}$

h) $P = \begin{bmatrix} -3 & 0 & 0 \\ 0 & 5 & 0 \\ 0 & 0 & 5 \end{bmatrix}, D = \begin{bmatrix} 2 & 1 & 0 \\ -1 & 0 & 1 \\ -2 & 3 & 4 \end{bmatrix}$

i) $P = \begin{bmatrix} 2 & 0 & 0 & 0 \\ 0 & 2 & 0 & 0 \\ 0 & 0 & 3 & 0 \\ 0 & 0 & 0 & 5 \end{bmatrix}, D = \begin{bmatrix} 3 & 0 & 3 & 1 \\ 0 & 3 & 2 & 0 \\ -2 & -1 & 0 & 0 \\ -1 & 1 & 0 & 0 \end{bmatrix}$

j) $P = \begin{bmatrix} 2 & 0 & 0 & 0 \\ 0 & 2 & 0 & 0 \\ 0 & 0 & 4 & 0 \\ 0 & 0 & 0 & 4 \end{bmatrix}, D = \begin{bmatrix} 0 & 0 & 2 & 0 \\ 0 & 0 & 0 & 1 \\ 1 & 0 & 0 & 0 \\ 0 & 1 & 1 & 0 \end{bmatrix}$

8 a) $\begin{bmatrix} 1 & 2 \\ -2 & 1 \end{bmatrix}$ b) $\begin{bmatrix} 2 & -1 \\ 0 & 2 \end{bmatrix}$

c) $\begin{bmatrix} -7 & -2 & -6 \\ 0 & -4 & -6 \\ 0 & 0 & -1 \end{bmatrix}$ d) $\begin{bmatrix} 8 & 3 & -6 \\ 0 & 1 & 3 \\ 0 & 0 & -3 \end{bmatrix}$

9 a) $b_1 = \begin{bmatrix} 1 \\ 1 \end{bmatrix}, \; b_2 = \begin{bmatrix} 1 \\ 3 \end{bmatrix}$

b) $b_1 = \begin{bmatrix} 3 \\ 7 \end{bmatrix}, \; b_2 = \begin{bmatrix} 1 \\ -1 \end{bmatrix}$

c) $b_= \begin{bmatrix} -2 \\ 1 \end{bmatrix}, \; b_2 = \begin{bmatrix} 1 \\ 1 \end{bmatrix}$

d) $b_1 = \begin{bmatrix} 3 \\ 1 \end{bmatrix}, \; b_2 = \begin{bmatrix} 2 \\ -1 \end{bmatrix}$

10 $b_1 = \begin{bmatrix} 11 \\ -3 \\ 4 \\ 4 \end{bmatrix}, b_2 = \begin{bmatrix} 0 \\ 1 \\ -1 \\ \frac{-2}{3} \end{bmatrix}, \; b_3 = \begin{bmatrix} 1 \\ 0 \\ \frac{5}{41} \\ \frac{7}{41} \end{bmatrix}, \; b_4 = \begin{bmatrix} 0 \\ 1 \\ \frac{-39}{41} \\ \frac{-30}{41} \end{bmatrix}$

Chapter 3

Orthogonality and Symmetric Matrices

3.1 Introduction

In this chapter, we introduce the concepts of distance, length and perpendicularity in a vector space $\mathbb{R}^n$. These geometric concepts provide powerful tools for solving many applied problems like Global Positioning System (GPS). All these geometric notions are defined in terms of an **inner product** **of two vectors in $\mathbb{R}^n$.**

3.2 Inner Product, Length and Orthogonality

Let $u, v \in \mathbb{R}^n$ as follows

$$u = \begin{bmatrix} u_1 \\ u_2 \\ \vdots \\ u_n \end{bmatrix} \quad \text{and} \quad v = \begin{bmatrix} v_1 \\ v_2 \\ \vdots \\ v_n \end{bmatrix}$$

Then the **innerproduct** of u and v is defined as

$$u^T v = [u_1, u_2 \cdots u_n] \begin{bmatrix} v_1 \\ v_2 \\ \vdots \\ v_n \end{bmatrix}$$

$$= u_1 v_1 + u_2 v_2 + \cdots u_n v_n$$

Note that $u^T v$ is a 1×1 matrix. Which we write as a single number (a scalar) without brackets. This **inner product** of two vectors is also called as a **dot product** of two vectors. It is written as $\mathbf{u} \cdot \mathbf{v}$.

Example 3.1. *Compute the inner product of $\mathbf{u}$ and $\mathbf{v}$ in $\mathbb{R}^3$, where*

$$\mathbf{u} = \begin{bmatrix} 2 \\ -5 \\ -1 \end{bmatrix} \quad \text{and} \quad \mathbf{v} = \begin{bmatrix} 3 \\ 2 \\ -3 \end{bmatrix}$$

$$u^T v = [2 - 5 - 1] \begin{bmatrix} 3 \\ 2 \\ -3 \end{bmatrix}$$

$$= 2 \times 3 + (-5) \times 2 + (-1)(-3)$$

$$= -1$$

Note that $\mathbf{u^T v = v^T u}$ (verify)

The following theorem states some important properties of an inner product of two vectors.

Theorem 3.1. *Let* $\mathbf{u, v, w} \in \mathbb{R}^n$, $\mathbf{c} \in \mathbb{R}$ *(c is a scalar belonging to* $\mathbb{R}$*) Then*

(a) $\mathbf{u \cdot v = v \cdot u}$ *(commutativity of inner product)*

(b) $\mathbf{(u + v) \cdot w = u \cdot w + v \cdot w}$ *(Distributive property)*

(c) $\mathbf{(cu) \cdot v} = c\mathbf{(u \cdot v)} = \mathbf{u \cdot (cv)}$ *(Associative property)*

(d) $\mathbf{u \cdot u} \geq 0$, *and* $\mathbf{u \cdot u} = 0$ *iff* $\mathbf{u = 0}$.   *(non-negativity property)*

Proof: Let $u = \begin{bmatrix} u_1 \\ u_2 \\ \vdots \\ u_n \end{bmatrix}$, $v = \begin{bmatrix} v_1 \\ v_2 \\ \vdots \\ v_n \end{bmatrix}$ and $w = \begin{bmatrix} w_1 \\ w_2 \\ \vdots \\ w_n \end{bmatrix}$ be given vectors in $\mathbb{R}^n$.

(a)

$$\mathbf{u \cdot v} = u_1 v_1 + u_2 v_2 + \cdots + u_n v_n \quad \text{(b(y definition)}$$

$$= v_1 u_1 + v_2 u_2 + \cdots + v_n u_n$$

$$= \mathbf{v \cdot u}$$

(b)

$$(\mathbf{u + v}) \cdot \mathbf{w} = [u_1 + v_1 \quad u_2 + v_2 \quad u_3 + v_3 \quad \cdots u_n + v_n] \begin{bmatrix} w_1 \\ w_2 \\ \vdots \\ w_n \end{bmatrix}$$

$$= (u_1 + v_1)w_1 + (u_2 + v_2)w_2 + (u_3 + v_3)w_3 \cdots + (u_n + v_n)w_n$$

$$= (u_1 w_1 + u_2 w_2 + \cdots + u_n w_n) + (v_1 w_1 + v_2 w_2 + \cdots + v_n w_n)$$

$$= \mathbf{u \cdot w + v \cdot w}$$

(c)

$$(\mathbf{cu}) \cdot \mathbf{v} = \begin{bmatrix} cu_1 & cu_2 & cu_3 & \cdots cu_n \end{bmatrix} \begin{bmatrix} v_1 \\ v_2 \\ \vdots \\ v_n \end{bmatrix}$$

$$= cu_1v_1 + cu_2v_2 + cu_3v_3 \cdots + cu_nv_n$$

$$= c(u_1v_1 + u_2v_2 + u_3v_3 \cdots + u_nv_n)$$

$$= c(\mathbf{u} \cdot \mathbf{v})$$

Similarly, $(\mathbf{cu}) \cdot \mathbf{v} = \mathbf{u} \cdot (\mathbf{cv})$ is proved.

(d) $\mathbf{u} \cdot \mathbf{u} = u_1^2 + u_2^2 + \cdots + u_n^2$.

As the above sum is a 'sum of squares of real values' $\mathbf{u} \cdot \mathbf{u}$ is always **non-negative**.

In particular, if $\mathbf{u} \cdot \mathbf{u} = 0$, this implies that each term $u_1^2, u_2^2, \cdots, u_n^2$ is zero. Hence $\mathbf{u} = \mathbf{0}$.

Conversely, if $\mathbf{u} = \mathbf{0}$ then $\mathbf{u} \cdot \mathbf{u} = 0$ is obvious. Hence proved the theorem.

Now we define **length** of a vector $\mathbf{u}$ in $\mathbb{R}^\mathbf{n}$, as follows:

Definition 3.1. *The* **length (or norm)** *of a vector* $\mathbf{u} \in \mathbb{R}^n$.

Let $\mathbf{u} \in \mathbb{R}^n$ *then length*

$$\mathbf{u} = \text{ norm } \mathbf{u} = \|\mathbf{u}\| = \sqrt{\mathbf{u} \cdot \mathbf{u}} = \sqrt{u_1^2 + u_2^2 + \cdots + u_n^2}.$$

As a consequence of these definition $\|\mathbf{u}\|^2 = \mathbf{u} \cdot \mathbf{u}$.

Geometric interpretation of $\|u\|$ **in** $\mathbb{R}^2$:

Let $\mathbf{u} = \begin{bmatrix} a \\ b \end{bmatrix}$, $a, b \in \mathbb{R}$, be a vector in $\mathbb{R}^2$. By definition of $\|\mathbf{u}\|$, $\|\mathbf{u}\| = \sqrt{a^2 + b^2}$

This value of $\|\mathbf{u}\|$ coincides with the standard notion of the length of the line segment from the origin to $\mathbf{u}$. (using Pythagoras Theorem)

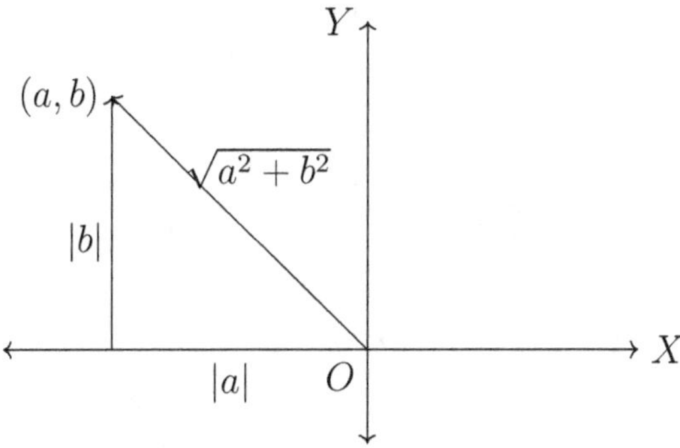

The notion of $\|\mathbf{u}\|$ of a vector $\mathbf{u} \in \mathbb{R}^3$ follows similarly as a length of the diagonal of a rectangular box. This coincides with the usual notion of length geometrically.

Similarly, the length of a given vector can be scaled as $\|\mathbf{cu}\| = |c|\|\mathbf{u}\|$ with an appropriate real number c.

Definition 3.2. *A **unit vector** in $\mathbb{R}^n$. A vector $\mathbf{u}$ in $\mathbb{R}^n$ called a **unit vector** if $\|\mathbf{u}\| = 1$.*

Given a vector $\mathbf{v} \in \mathbb{R}^n$, such that $\|\mathbf{v}\| \neq 1$ then we can find the vector corresponding to $\mathbf{v}$ such that its length is 1. This is achieved by dividing the given vector $\mathbf{v}$ by its length (norm), that is $\|\mathbf{v}\|$. That is, to be precise, for any vector $\mathbf{v} \in \mathbb{R}^n$ the corresponding unit vector is $\frac{\mathbf{v}}{\|\mathbf{v}\|}$. This process of finding a unit vector from a given vector in $\mathbb{R}^n$ is called **normalization of a vector.** This unit vector so obtained is in the same direction as vector $\mathbf{v}$.

The following examples illustrate the normalization notion.

Example 3.2. *Find a **unit vector** in the direction of a given vector*

$$(a)\ \mathbf{u} = \begin{bmatrix} 7 \\ 4 \\ \frac{1}{2} \\ 1 \end{bmatrix} \qquad (b)\ \mathbf{v} = \begin{bmatrix} \frac{8}{3} \\ 2 \end{bmatrix}.$$

Solution:

(a) Here

$$\mathbf{u} = \begin{bmatrix} \frac{7}{4} \\ \frac{1}{2} \\ 1 \end{bmatrix}$$

$$\|u\| = \sqrt{\frac{49}{16} + \frac{1}{4} + 1} = \frac{\sqrt{69}}{4}$$

Therefore, $\frac{\mathbf{u}}{\|\mathbf{u}\|} = \begin{bmatrix} \frac{7}{\sqrt{69}} \\ \frac{2}{\sqrt{69}} \\ \frac{4}{\sqrt{69}} \end{bmatrix}$ is a unit vector.

(b) Here

$$\mathbf{v} = \begin{bmatrix} \frac{8}{3} \\ 2 \end{bmatrix}$$

$$\|v\| = \sqrt{\frac{64}{9} + 4} = \frac{10}{3}$$

Therefore, $\frac{\mathbf{v}}{\|\mathbf{v}\|} = \begin{bmatrix} \frac{4}{5} \\ \frac{3}{5} \end{bmatrix}$ is a unit vector.

Geometrically, the above unit vector are as follows:

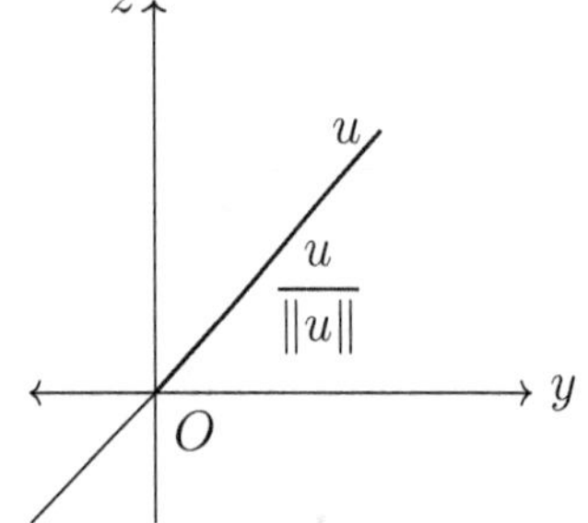

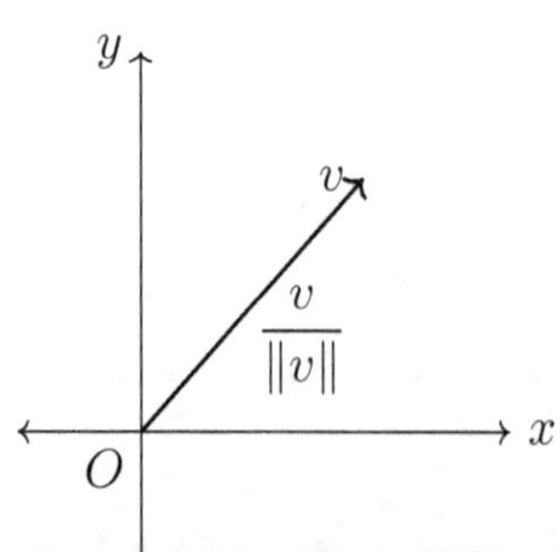

At this point we define distance between two vectors in $\mathbb{R}^n$.

In $\mathbb{R}$, we know that if $a, b \in \mathbb{R}$ then distance between a and b is defined as $|a - b|$. This definition of distance in $\mathbb{R}$ has a direct analogue in $\mathbb{R}^n$.

Definition 3.3. Distacne between two vectors in $\mathbb{R}^n$.

For $\mathbf{u}, \mathbf{v} \in \mathbb{R}^n$, *dist* $(\mathbf{u}, \mathbf{v})$ *is the length of the vector* $\mathbf{u} - \mathbf{v}$.

$$dist(\mathbf{u}, \mathbf{v}) = \|u - v\|.$$

Example 3.3. *Compute the following distances between given vectors:*

(a) $\mathbf{x} = \begin{bmatrix} 10 \\ -3 \end{bmatrix}$, $\mathbf{y} = \begin{bmatrix} -1 \\ -5 \end{bmatrix}$ *(b)* $\mathbf{u} = \begin{bmatrix} 0 \\ -5 \\ 2 \end{bmatrix}$, $\mathbf{v} = \begin{bmatrix} -4 \\ -1 \\ 8 \end{bmatrix}$.

Solution: (a) The distance between the given vectors is

$$\text{dist} (\mathbf{x}, \mathbf{y}) = \|\mathbf{x} - \mathbf{y}\|$$

$$= \sqrt{11^2 + 2^2} = \sqrt{121 + 4} = \sqrt{125} = 5\sqrt{5} \quad \text{units}$$

(b) The distance between the given vectors obtain as follows

$$\mathbf{u} = \begin{bmatrix} 0 \\ -5 \\ 2 \end{bmatrix}, \; \mathbf{v} = \begin{bmatrix} -4 \\ -1 \\ 8 \end{bmatrix}$$

$$\therefore \mathbf{u} - \mathbf{v} = \begin{bmatrix} 0 \\ -5 \\ 2 \end{bmatrix} - \begin{bmatrix} -4 \\ -1 \\ 8 \end{bmatrix} = \begin{bmatrix} 4 \\ -4 \\ -6 \end{bmatrix}$$

$$\therefore \|\mathbf{u} - \mathbf{v}\| = \sqrt{4^2 + (-4)^2 + (-6)^2}$$

$$= \sqrt{16 + 16 + 36}$$

$$= \sqrt{68}$$

$$= 2\sqrt{17} \quad \text{units}$$

Example 3.4. *If* $\mathbf{w} = \begin{bmatrix} 3 \\ -1 \\ -5 \end{bmatrix}$, $\mathbf{x} = \begin{bmatrix} 6 \\ -2 \\ 3 \end{bmatrix}$ *then compute the quantity* $\left(\frac{\mathbf{x} \cdot \mathbf{w}}{\mathbf{x} \cdot \mathbf{x}}\right) \mathbf{x}$.

Solution: We have

$$\mathbf{x} \cdot \mathbf{w} = 3 \times 6 + (-1) \times (-2) + (-5) \times (3)$$

$$= 5$$

$$\mathbf{x} \cdot \mathbf{x} = 3 \times 3 + (-1) \times (-1) + (-5) \times (-5)$$

$$= 35$$

$$\text{Therefore,} \quad \left(\frac{\mathbf{x} \cdot \mathbf{w}}{\mathbf{x} \cdot \mathbf{x}}\right) \mathbf{x} = \left(\frac{5}{35}\right) \begin{bmatrix} 6 \\ -2 \\ 3 \end{bmatrix} = \frac{1}{7} \begin{bmatrix} 6 \\ -2 \\ 3 \end{bmatrix}$$

The notion of orthogonal vectors in $\mathbb{R}^n$ plays an important role. In $\mathbb{R}^2$, two vectors $\mathbf{u}$ and $\mathbf{v}$ are orthogonal to each other whenever they are perpendicular to each other in Euclidean Geometry. This concept of perpendicular lines in $\mathbb{R}^2$ is looked upon in terms of distance (hence in terms of norm) as follows. In the following figure **the two lines through origin determined by vectors u and v are geometrically perpendicular iff the distance from u to v is same as the distance from u to $-v$.**

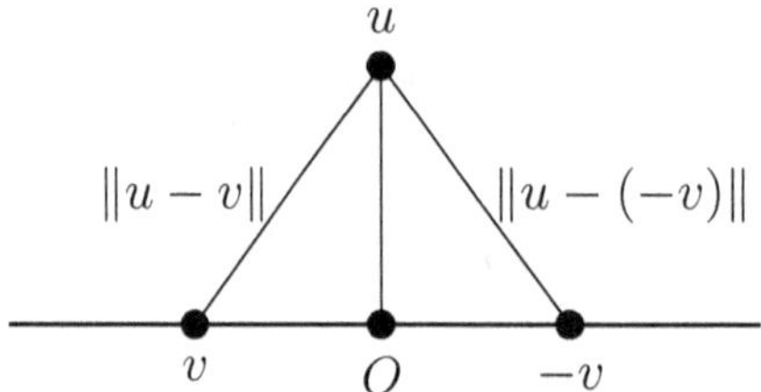

We compute the said distances as follows:

$$\|u - v\|^2 = (u - v) \cdot (u - v)$$
$$= u \cdot (u - v) - v \cdot (u - v)$$
$$= (u \cdot u) - u \cdot v - v \cdot u + v \cdot v$$
$$= \|u\|^2 - 2u \cdot v + \|v\|^2 \tag{3.1}$$

Similarly,

$$\|u - (-v)\|^2 = \|u + v\|^2$$
$$= \|u\|^2 + 2u \cdot v + \|v\|^2 \tag{3.2}$$

The equations (3.1) and (3.2) are equal iff $2u \cdot v = -2u \cdot v$

This is true iff $u \cdot v = 0$. This is the require condition of orthogonality. We generalize it to $\mathbb{R}^n$.

Definition 3.4. Orthogonal vectors in $\mathbb{R}^n$

Two vectors $\mathbf{u}$ and $\mathbf{v}$ in $\mathbb{R}^n$ are orthogonal (to each other) if and only if $\mathbf{u} \cdot \mathbf{v} = \mathbf{0}$.

Note that a **zero vector** (0) is orthogonal to every other vector $\mathbf{v} \in \mathbb{R}^n$.

From equations (3.1), (3.2) and definition of orthogonality of vectors in $\mathbb{R}^n$ the following theorem follows.

Theorem 3.2. *Two vectors $\mathbf{u}$ and $\mathbf{v}$ are orthogonal iff $\|\mathbf{u} + \mathbf{v}\|^2 = \|\mathbf{u}\|^2 + \|\mathbf{v}\|^2$.*

See figure below that explain statement of theorem $3 \cdot 2$ geometrically.

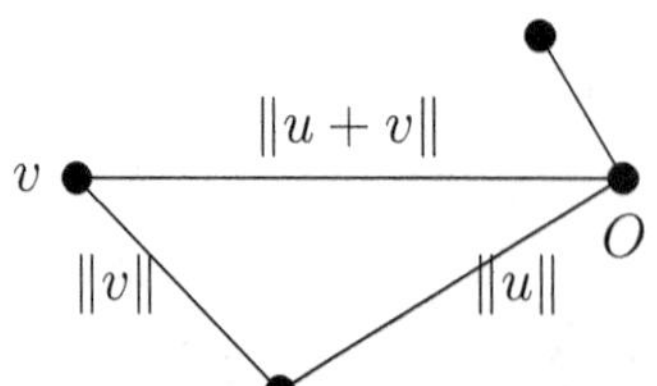

Example 3.5. *Determine if the given pair of vectors is orthogonal or not*

$$\mathbf{u} = \begin{bmatrix} 2 \\ 2 \\ -5 \\ 0 \end{bmatrix}, \quad \mathbf{v} = \begin{bmatrix} -4 \\ 1 \\ -2 \\ 6 \end{bmatrix}.$$

Solution: Note that

$$\mathbf{u} \cdot \mathbf{v} = 3 \times (-4) + (2)(1) + (-5)(-2) + 0(6)$$

$$= -12 + 2 + 10 + 0$$

$$= 0$$

Hence $\mathbf{u}$ and $\mathbf{v}$ are orthogonal to each other.

Example 3.6. *Show that the vectors $\mathbf{u}$ and $\mathbf{v}$ are orthogonal to each other*

$$\mathbf{u} = \begin{bmatrix} 12 \\ 3 \\ -5 \end{bmatrix}, \quad \mathbf{v} = \begin{bmatrix} 2 \\ -3 \\ 3 \end{bmatrix}.$$

Solution: We show orthogonality we use theorem (3.2).

$$\|\mathbf{u} + \mathbf{v}\|^2 = 14^2 + (-2)^2 = 200 \tag{3.3}$$

$$\|u\|^2 = 12^2 + 3^2 + (-5)^2 = 178$$

$$\|v\|^2 = 2^2 + (-3)^2 + 3^2 = 22$$

$$\therefore \|u\|^2 + \|v\|^2 = 178 + 22 = 200. \tag{3.4}$$

Therefor from equations (3.3) and (3.4), we obtain

$$\|u + v\|^2 = \|u\|^2 + \|v\|^2.$$

Hence, theorem (3.2) is satisfied.

Therefore,, vectors $\mathbf{u}$ and $\mathbf{v}$ are orthogonal to each other.

Using inner product of two vectors $\mathbf{u}$ and $\mathbf{v}$ in $\mathbb{R}^n$ we define angle between them as follows:

Definition 3.5. Angle between two vectors in $\mathbb{R}^n$.

If $\mathbf{u}$ and $\mathbf{v}$ are non-zero vectors in $\mathbb{R}^n$ then the angle θ between $\mathbf{u}$ and $\mathbf{v}$ is defined as

$$\cos \theta = \frac{\mathbf{u} \cdot \mathbf{v}}{\|\mathbf{u}\| \, \|\mathbf{v}\|}.$$

Example 3.7. *Find angle between the following vectors*

$$\mathbf{u} = \begin{bmatrix} 1 \\ 0 \end{bmatrix}, \quad \mathbf{v} = \begin{bmatrix} -1 \\ 1 \end{bmatrix}$$

Solution: Here $\mathbf{u} \cdot \mathbf{v} = -2$

$$\|u\| = \sqrt{1^2 + 1^2} = \sqrt{2}, \ \|v\| = \sqrt{3}$$

$$\cos \theta = \frac{u \cdot v}{\|u\| \ \|v\|}$$

$$= \frac{-2}{\sqrt{2} \cdot \sqrt{3}} = -\sqrt{\frac{2}{3}}$$

$$\therefore \theta = \cos^{-1}\left(-\sqrt{\frac{2}{3}}\right)$$

3.3 Orthogonal Sets

We have seen the notion of orthogonality between two vectors of $\mathbb{R}^n$. Now, extending the idea further, we define orthogonal set of vectors as follows.

Definition 3.6. An orthogonal set

A set of vectors $\{u_1, u_2, \cdots, u_k\} \in \mathbb{R}^n$ is said to be an **orthogonal set,** *if each pair of distinct vectors from the set is orthogonal, that is $u_j \cdot u_i = 0, \ \forall \ i \neq j$*

Example 3.8. *Determine whether the following sets of vectors are orthogonal or not.*

Solution: (a) $u_1 = \begin{bmatrix} 1 \\ -2 \\ 1 \end{bmatrix}$, $u_2 = \begin{bmatrix} 0 \\ 1 \\ 2 \end{bmatrix}$, $u_3 = \begin{bmatrix} -5 \\ -2 \\ 1 \end{bmatrix}$

We see that, $u_1 \cdot u_2 = 0, \ u_2 \cdot u_3 = 0$ and $u_1 \cdot u_3 = 0$.

Hence the given set of non-zero vectors $\{u_1, u_2, u_3\}$ is an orthogonal set.

(b) $u_1 = \begin{bmatrix} -1 \\ 4 \\ -3 \end{bmatrix}$, $u_2 = \begin{bmatrix} 5 \\ 2 \\ 1 \end{bmatrix}$, $u_3 = \begin{bmatrix} 3 \\ -4 \\ -7 \end{bmatrix}$

Here, we see that $u_1 \cdot u_3 = 2 \neq 0$ hence the given set is not orthogonal.

The following theorem tells us an additional property of an orthogonal set of non-zero vectors of $\mathbb{R}^n$.

Theorem 3.3. *If $S = \{u_1, u_2, \cdots, u_k\}$ is an orthogonal set of non-zero vectors in $\mathbb{R}^n$, then S is linearly independent and hence is a basis for the subspace spanned by S.*

Proof: First we show that the set S is linearly independent.

Suppose, $c_1 u_1 + c_2 u_2 + \cdots + c_k u_k = 0$ where $c_1, c_2 \cdots, c_k$ are scalars.

$$(c_1 u_1 + c_2 u_2 + \cdots + c_k u_k) \cdot u_1 = 0 \cdot u_1$$

$$\therefore c_1(u_1 \cdot u_1) + c_2(u_2 \cdot u_1) + \cdots + c_k(u_k \cdot u_1) = 0.$$

$$\therefore c_1(u_1 \cdot u_1) = 0.$$

But u_1 is a non-zero vector in $\mathbb{R}^n$.

So, S is a set that spans a subspace and S is linearly independent.

Therefore, it forms a basis of the subspace spanned by set S.

Hence, set S forms an orthogonal basis.

Definition 3.7. An orthogonal basis

An orthogonal basis for a subspace W of $\mathbb{R}^n$ is a basis for W which is also an orthogonal set.

The weights (coefficients) in the linear combination of vectors of a basis are easily computed if the basis if orthogonal. So orthogonal basis is preferred than other bases in some situations. The following theorem suggest the said advantage.

Theorem 3.4. *Let W be a subspace of $\mathbb{R}^n$ with an orthogonal basis $S = \{u_1, u_2, \cdots, u_k\}$. For each $y \in W$ the weights (coefficients) in the linear combination $y = c_1 u_1 + c_2 u_2 + \cdots + c_k u_k$ are given by*

$$c_i = \frac{y \cdot u_i}{u_i \cdot u_i}, \quad i = 1, 2, \cdots k.$$

Proof: S is an orthogonal set.

Therefore,

$$y \cdot u_1 = c_1 u_1 \cdot u_1 + c_2 u_2 \cdot u_1 + \cdots + c_k u_k \cdot u_1$$

$$= c_1(u_1 \cdot u_1) + 0 + 0 + \cdots + 0$$

$$y \cdot u_1 = c_1(u_1 \cdot u_1)$$

$$\therefore c_1 = \frac{y \cdot u_1}{u_1 \cdot u_1} \tag{3.5}$$

The equation (3.5) is true for all $i = 1, 2, \cdots k$.

Hence, $c_i = \dfrac{y \cdot u_i}{u_i \cdot u_i}$ for any i with $u_i \cdot u_i \neq 0 \;\forall_i = 1, 2, \cdots k$.

Hence, proved the theorem.

Example 3.9. *Let $S = \{u_1, u_2, u_3\}$ be an orthogonal basis of $\mathbb{R}^3$. Express x as a linear combination of $u_{i'}s$.*

Solution: From theorem (3.4) we compute the weights (coefficients) as follows.

Given $u_1 = \begin{bmatrix} 1 \\ 0 \\ 1 \end{bmatrix}$, $u_2 = \begin{bmatrix} -1 \\ 4 \\ 1 \end{bmatrix}$, $u_3 = \begin{bmatrix} 2 \\ 1 \\ -2 \end{bmatrix}$, $x = \begin{bmatrix} 8 \\ -4 \\ -3 \end{bmatrix}$.

For the linear combination $c_1 u_1 + c_2 u_2 + c_3 u_3 = x$

$$c_1 = \frac{x \cdot u_1}{u_1 \cdot u_1} = \frac{5}{2}$$

$$c_2 = \frac{x \cdot u_2}{u_2 \cdot u_2} = \frac{-27}{18} = \frac{-9}{6} = \frac{-3}{2}$$

$$c_3 = \frac{x \cdot u_3}{u_3 \cdot u_3} = \frac{18}{9} = 2$$

$$\frac{5}{} \qquad \frac{3}{} \qquad + 2$$

3.3.1   Orthogonal Projections

In this section, we express a vector $y \in \mathbb{R}^n$ as a sum of two vectors in $\mathbb{R}^n$. Out of the two vectors one is given as vector $u \in \mathbb{R}^n$. We find other vector 'z' orthogonal to 'u'. That is,

$$y = \alpha.u + z = \hat{y} + z$$

for some scalar α and z is a vector orthogonal to vector 'u'. Geometrically it is shown in the following figure.

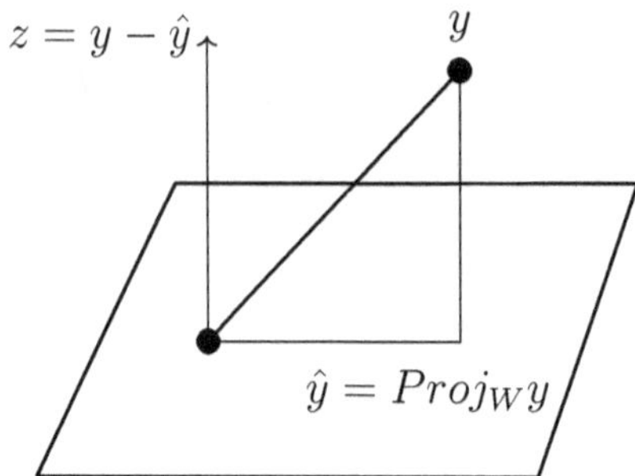

Given any scalar α, Let $\mathbf{z} = \mathbf{u} - \alpha\mathbf{u}$. So that equation $\mathbf{y} = \alpha\mathbf{u} + \mathbf{z}$ is satisfied. Then $\mathbf{y} - \alpha\mathbf{u}$ is orthogonal to $\mathbf{u}$ iff $0 = (y - \alpha u) \cdot u = y \cdot u - \alpha u \cdot u$

That is, the equation $y = \alpha u + z$ **is satisfied iff** $\alpha = \dfrac{y \cdot u}{u \cdot u}$ **and** $\hat{y} = \dfrac{y \cdot u}{u \cdot u} u.$

The vector $\hat{\mathbf{y}}$ is called **orthogonal projection of** y **onto** u. The vector $\mathbf{z}$ is known as the component of $\mathbf{y}$ orthogonal to $\mathbf{u}$. This projection is independent of any non-zero multiple of vector $\mathbf{u}$. Hence this projection is determined by the subspace $\mathbf{L}$ spanned by $\mathbf{u}$. We get the following formula of the **orthogonal projection of** y **onto space** L. It is defined as follows:

$$\hat{y} = \text{proj}_L y = \frac{y \cdot u}{u \cdot u} u$$

The following example illustrates the above defined notion of projection.

Example 3.10. *Find the orthogonal projection of vector* $\mathbf{y} = \begin{bmatrix} 7 \\ 6 \end{bmatrix}$ *onto vector* $\mathbf{u} = \begin{bmatrix} 4 \\ 2 \end{bmatrix}$. *Then write y as the sum of two orthogonal vectors $\hat{y}$ and z.*

Solution: We know that

$$\hat{y} = \left(\frac{y \cdot u}{u \cdot u} \right) u = \frac{\begin{bmatrix} 7 \\ 6 \end{bmatrix} \cdot \begin{bmatrix} 4 \\ 2 \end{bmatrix}}{\begin{bmatrix} 4 \\ 2 \end{bmatrix} \cdot \begin{bmatrix} 4 \\ 2 \end{bmatrix}} = \frac{40}{20} = 2u$$

and $\mathbf{y} - \hat{\mathbf{y}}$ is the component of $\mathbf{y}$ orthogonal to $\mathbf{u}$ defined by $\mathbf{z}$.

$$\mathbf{z} = \mathbf{y} - \hat{\mathbf{y}} = \begin{bmatrix} 7 \\ 6 \end{bmatrix} - 2 \begin{bmatrix} 4 \\ 2 \end{bmatrix}$$

$$= \begin{bmatrix} -1 \\ 2 \end{bmatrix}$$

$$\therefore y = \hat{y} + z = \begin{bmatrix} 8 \\ 4 \end{bmatrix} + \begin{bmatrix} -1 \\ 2 \end{bmatrix} = \mathbf{2u} + \mathbf{z}$$

Here we also verify that vectors $\hat{y}$ and $(y - \hat{y})$ are orthogonal components of y.

$$\hat{y} \cdot (y - \hat{y}) = \begin{bmatrix} 8 \\ 4 \end{bmatrix} \cdot \begin{bmatrix} -1 \\ 2 \end{bmatrix} = 0$$

The following graph shows the components geometrically

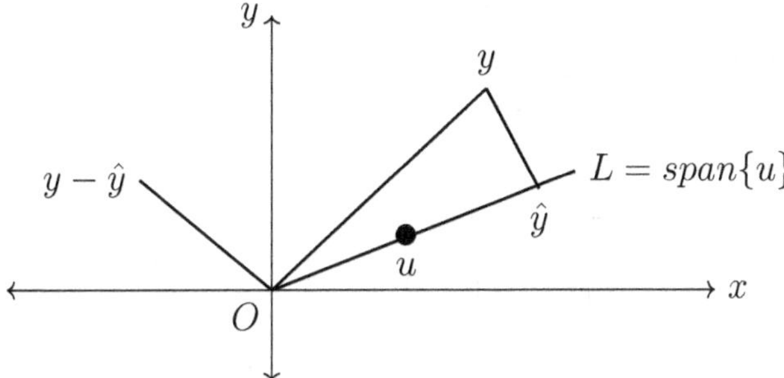

The length of the perpendicular from y onto L is $\|y - \hat{y}\| = \sqrt{(-1)^2 + 2^2} = \sqrt{5}$.

In theorem (3.4) We have seen that any vector $y \in W$ can be written as

$$y = \frac{y \cdot u_1}{u_1 \cdot u_1} u_1 + \frac{y \cdot u_2}{u_2 \cdot u_2} u_2 + \cdots + \frac{y \cdot u_k}{u_k \cdot u_k} u_k$$

where $\{u_1, u_2, \cdots, u_k\}$ is an orthogonal basis of W.

Now we can see that the first term in the above equation is the projection of y on to the subspace spanned by u_1 (it is the line through u_1 and origin), second term of it is the projection of y onto the subspace spanned by u_2. So the equation expresses y as a sum of its projections onto axes (orthogonal) determined by $u_1, u_2, \cdots, u_k$. These are one dimensional subspaces $u_1, u_2 \cdots u_k$ and are mutually orthogonal.

Definition 3.8. Orthonormal Set

*A set $\{u_1, u_2, \cdots, u_k\}$ is an **orthonormal** set if it is an orthogonal set of unit vectors.*

A space W spanned by an orthonormal set is said to have an **orthonormal basis** of W.

The standard basis $\{\mathbf{e_1}, \mathbf{e_2}, \cdots, \mathbf{e_n}\}$ of unit vectors of $\mathbb{R}^n$ is an example of orthonormal basis.

Any non-empty subset of an orthonormal set is orthonormal.

Given an orthogonal set of non-zero vectors, by normalizing the vectors we obtain vectors of unit

Example 3.11. *Determine whether the given set of vectors is orthonormal or not.*

$$\mathbf{v_1} = \begin{bmatrix} \frac{1}{3} \\ \frac{1}{3} \\ \frac{1}{3} \end{bmatrix}, \quad \mathbf{v_2} = \begin{bmatrix} \frac{-1}{2} \\ 0 \\ \frac{1}{2} \end{bmatrix}$$

Solution: We obtain for given vectors $\mathbf{v_1} \cdot \mathbf{v_2} = \frac{-1}{6} + \frac{1}{6} = 0$

So, $\mathbf{v_1}$ and $\mathbf{v_2}$ are orthogonal to each other.

Now, we check whether $\mathbf{v_1}, \mathbf{v_2}$ are unit vectors or not.

$$v_1 \cdot v_1 = \frac{1}{9} + \frac{1}{9} + \frac{1}{9} = \frac{3}{9} = \frac{1}{3} \neq 1$$
$$v_2 \cdot v_2 = \frac{1}{4} + 0 + \frac{1}{4} = \frac{2}{4} = \frac{1}{2} \neq 1$$

So the given vectors are not unit vectors. In order to convert them to unit vectors we need to divide v_1 by $\|v_1\|$ and v_2 by $\|v_2\|$ as follows:

$$\|v_1\| = \sqrt{\frac{1}{9} + \frac{1}{9} + \frac{1}{9}} = \sqrt{\frac{1}{3}},$$
$$\|v_2\| = \sqrt{\frac{1}{4} + 0 + \frac{1}{4}} = \sqrt{\frac{1}{2}}$$

$$\therefore w_1 = \frac{v_1}{\|v_1\|} = \frac{v_1}{\sqrt{\frac{1}{3}}} = \frac{1}{\sqrt{\frac{1}{3}}} \begin{bmatrix} \frac{1}{3} \\ \frac{1}{3} \\ \frac{1}{3} \end{bmatrix} = \begin{bmatrix} \frac{1}{\sqrt{3}} \\ \frac{1}{\sqrt{3}} \\ \frac{1}{\sqrt{3}} \end{bmatrix}$$

$$w_2 = \frac{v_2}{\|v_2\|} = \frac{1}{\sqrt{\frac{1}{2}}} \begin{bmatrix} \frac{-1}{2} \\ 0 \\ \frac{1}{2} \end{bmatrix} = \begin{bmatrix} \frac{-1}{\sqrt{2}} \\ 0 \\ \frac{1}{\sqrt{3}} \end{bmatrix}$$

The set $\{w_1, w_2\}$ is orthonormal set.

Example 3.12. *Given* $u_1 = \begin{bmatrix} \frac{1}{\sqrt{10}} \\ \frac{3}{\sqrt{20}} \\ \frac{3}{\sqrt{20}} \end{bmatrix}, \quad u_2 = \begin{bmatrix} \frac{3}{\sqrt{10}} \\ \frac{-1}{\sqrt{20}} \\ \frac{-1}{\sqrt{20}} \end{bmatrix}, \quad u_3 = \begin{bmatrix} 0 \\ \frac{-1}{\sqrt{2}} \\ \frac{1}{\sqrt{2}} \end{bmatrix}$

Determine whether the set $\{u_1, u_2, u_3\}$ *is an orthonormal set. If the set is orthogonal, normalize the vectors to produce an orthonormal set.*

Solution: First we check orthogonality of the given vectors.

$$u_1 \cdot u_2 = \frac{3}{10} + \frac{-3}{20} + \frac{-3}{20} = 0$$
$$u_2 \cdot u_3 = 0 + \frac{1}{\sqrt{40}} - \frac{1}{\sqrt{40}} = 0$$
$$u_1 \cdot u_3 = 0 + \frac{-3}{\sqrt{40}} + \frac{3}{\sqrt{40}} = 0$$

This shows that $\{u_1, u_2, u_3\}$ is orthogonal set. Now we test orthonormality of the vectors as follows:

$$u_1 \cdot u_1 = \frac{1}{10} + \frac{9}{20} + \frac{9}{20} = \frac{20}{20} = 1$$

$$u_2 \cdot u_2 = \frac{9}{10} + \frac{1}{20} + \frac{1}{20} = \frac{20}{20} = 1$$

$$u_3 \cdot u_3 = 0 + \frac{1}{2} + \frac{1}{2} = \frac{2}{2} = 1$$

which shows that u_1, u_2, u_3 are unit vectors. Thus the set $\{u_1, u_2, u_3\}$ is an orthonormal set.

The orthonormal sets are important in applications and in computer algorithms for matrix computations.

A matrix of order $m \times n$ whose column vectors form an orthonormal set play important role in computer science. The following theorem are useful for matrices with orthonormal columns.

Theorem 3.5. *An $m \times n$ matrix U has orthonormal columns iff $U^T U = I$.*

Theorem 3.6. *Let U be an $m \times n$ matrix with orthonormal columns and let $\mathbf{x}, \mathbf{y} \in \mathbb{R}^n$, Then*

(a) $\|Ux\| = \|x\|$

(b) $(Ux) \cdot (Uy) = x \cdot y$

(c) $(Ux) \cdot (Uy) = 0$ iff $x \cdot y = 0$.

From (a) and (c) it shows that the linear map $x \longmapsto Ux$ preserves length and orthogonality of vectors.

Example 3.13. *Let* $\mathbf{U} = \begin{bmatrix} \frac{1}{\sqrt{2}} & \frac{2}{3} \\ \frac{1}{\sqrt{2}} & \frac{-2}{3} \\ 0 & \frac{1}{3} \end{bmatrix}$ *and* $\mathbf{x} = \begin{bmatrix} \sqrt{2} \\ 3 \end{bmatrix}$. *Verify theorem (3.6) for* $\mathbf{U}$ *and* $\mathbf{x}$.

Solution: First of all we show that columns of U are orthonormal by applying theorem (3.5).

$$U^T U = \begin{bmatrix} \frac{1}{\sqrt{2}} & \frac{1}{\sqrt{2}} & 0 \\ \frac{2}{3} & \frac{-2}{3} & \frac{1}{3} \end{bmatrix} \begin{bmatrix} \frac{1}{\sqrt{2}} & \frac{2}{3} \\ \frac{1}{\sqrt{2}} & \frac{-2}{3} \\ 0 & \frac{1}{3} \end{bmatrix} = \begin{bmatrix} 1 & 0 \\ 0 & 1 \end{bmatrix}$$

So, U is a matrix with orthonormal columns.

$$Ux = \begin{bmatrix} 3 \\ -1 \\ 1 \end{bmatrix}$$

$$\therefore \|Ux\| = \sqrt{9 + 1 + 1} = \sqrt{11}$$

$$\text{and } \|x\| = \sqrt{2 + 9} = \sqrt{11}$$

Property (a) of theorem is verified.

Let $\mathbf{y} = \begin{bmatrix} -3\sqrt{2} \\ 6 \end{bmatrix}$.

Then

$$Ux = \begin{bmatrix} 3 \\ -1 \\ 1 \end{bmatrix}, \; Uy = \begin{bmatrix} 1 \\ -7 \\ 2 \end{bmatrix}$$

$$\therefore Ux \cdot Uy = 3 \times 1 + (-1) \times (-7) + 1 \times 2$$

$$= 3 + 7 + 2 = 12$$

$$\text{and } x \cdot y = \begin{bmatrix} \sqrt{2} \\ 3 \end{bmatrix} \cdot \begin{bmatrix} -3\sqrt{2} \\ 6 \end{bmatrix}$$

$$= \sqrt{2} \times (-3\sqrt{2}) + 3 \times 6$$

$$= -6 + 18 = 12$$

$$\therefore (Ux) \cdot (Uy) = x.y$$

Hence, (b) is verified.

Definition 3.9. An orthogonal Matrix

A square invertible matrix U such that $U^{-1} = U^T$ is called **an orthogonal matrix.**

In fact, it is easy to see that any square matrix with orthonormal columns is an orthogonal matrix. For example,

$$U = \begin{bmatrix} \frac{3}{\sqrt{11}} & \frac{-1}{\sqrt{6}} & \frac{-1}{\sqrt{66}} \\ \frac{1}{\sqrt{11}} & \frac{2}{\sqrt{6}} & \frac{-4}{\sqrt{66}} \\ \frac{1}{\sqrt{11}} & \frac{1}{\sqrt{6}} & \frac{7}{\sqrt{66}} \end{bmatrix}$$

3.4 Orthogonal Decomposition

In this section, we combine theorem (3.4) with the notion of orthogonal projection described in section $3 \cdot 2 \cdot 1$. As a result we state the following **Orthogonal Decomposition Theorem.**

Theorem 3.7. *The Orthogonal Decomposition Theorem*

Let W be a subspace of $\mathbb{R}^n$, let $\{u_1, u_2, \cdots, u_k\}$ be any orthogonal basis of W. Then each $y \in \mathbb{R}^n$ can be uniquely written in the form $y = \hat{y} + z$ where $\hat{y} \in W$ and $z = y - \hat{y}$, in fact

$$\hat{\mathbf{y}} = \frac{\mathbf{y} \cdot \mathbf{u_1}}{\mathbf{u_1} \cdot \mathbf{u_1}} \mathbf{u_1} + \frac{\mathbf{y} \cdot \mathbf{u_2}}{\mathbf{u_2} \cdot \mathbf{u_2}} \mathbf{u_2} + \cdots + \frac{\mathbf{y} \cdot \mathbf{u_k}}{\mathbf{u_k} \cdot \mathbf{u_k}} \mathbf{u_k}$$

The vector $\hat{y}$ is known as **the orthogonal projection of y on to W.** $\hat{y}$ *is written as* $\text{proj}_W y$.

The **uniqueness** *of the decomposition of* **vector** *y shows that the* **orthogonal projection** *$\hat{y}$ depends only on W and not on a particular orthogonal basis used.*

Geometric Interpretation of Orthogonal Projection :

If dimension of the subspace W is greates than **1** then each term in

$$\mathbf{y} \cdot \mathbf{u_1} \qquad \mathbf{y} \cdot \mathbf{u_2} \qquad \mathbf{y} \cdot \mathbf{u_k}$$

is itself an orthogonal projection of $\mathbf{y}$ onto a 1-dimensional subspace spanned by the $\mathbf{u}$'s in the basis of $\mathbf{W}$. The following figure explains geometrically this situation for a subspace W of $\mathbb{R}^3$ spanned by $\{\mathbf{u_1}, \mathbf{u_2}\}$.

We can see from the figure that, the orthogonal projection on $\mathbf{y}$ is the sum of its projections onto one-dimensional subspaces that are mutually orthogonal.

In the figure, $\mathbf{\hat{y}_1}$ and $\mathbf{\hat{y}_2}$ denote the projections of y onto the lines spanned by $\mathbf{u_1}$ and $\mathbf{u_2}$ respectively. Thus orthogonal projection of $\mathbf{y}$ as $\mathbf{\hat{y}}$ is the sum of the projections of $\mathbf{y}$ onto **one dimensional subspaces** that are orthogonal to each other.

Example 3.14. *Let* $\mathbf{W} = \text{span } \{\mathbf{u_1}, \mathbf{u_2}\}$ *be a subspace with an orthogonal basis.*

Let $\mathbf{u_1} = \begin{bmatrix} 2 \\ 5 \\ -1 \end{bmatrix}$, $u_2 = \begin{bmatrix} -2 \\ 1 \\ 1 \end{bmatrix}$ *and* $\mathbf{y} = \begin{bmatrix} 1 \\ 2 \\ 3 \end{bmatrix}$. *Write vector* $\mathbf{y}$ *as* $\mathbf{\hat{y}} + \mathbf{z}$.

Solution: We know that $\mathbf{\hat{y}}$ is an orthogonal projection of $\mathbf{y}$ onto $\mathbf{w}$.

$$\mathbf{\hat{y}} = \frac{\mathbf{y} \cdot \mathbf{u_1}}{\mathbf{u_1} \cdot \mathbf{u_1}} \mathbf{u_1} + \frac{\mathbf{y} \cdot \mathbf{u_2}}{\mathbf{u_2} \cdot \mathbf{u_2}} \mathbf{u_2}$$

$$\mathbf{\hat{y}} = \frac{9}{30} \begin{bmatrix} 2 \\ 5 \\ -1 \end{bmatrix} + \frac{3}{6} \begin{bmatrix} -2 \\ 1 \\ 1 \end{bmatrix} = \frac{9}{30} \begin{bmatrix} 2 \\ 5 \\ -1 \end{bmatrix} + \frac{5}{30} \begin{bmatrix} -2 \\ 1 \\ 1 \end{bmatrix} = \begin{bmatrix} \frac{-2}{5} \\ 2 \\ \frac{1}{5} \end{bmatrix}$$

$$\text{and } \mathbf{y} - \mathbf{\hat{y}} = \begin{bmatrix} 1 \\ 2 \\ 3 \end{bmatrix} - \begin{bmatrix} \frac{-2}{5} \\ 2 \\ \frac{1}{5} \end{bmatrix} = \begin{bmatrix} \frac{7}{5} \\ 0 \\ \frac{14}{5} \end{bmatrix}$$

Hence,

$$\mathbf{y} = \mathbf{\hat{y}} + (y - \hat{y})$$

$$\mathbf{y} = \begin{bmatrix} \frac{-2}{5} \\ 2 \\ \frac{1}{5} \end{bmatrix} + \begin{bmatrix} \frac{7}{5} \\ 0 \\ \frac{14}{5} \end{bmatrix}$$

This is the required decomposition of $\mathbf{y}$.

The vector $\hat{y}$ so defined in theorem $3 \cdot 6$ is called the **best approximation for vector** y by **elements of** W.

The following theorem state this fact.

Theorem 3.8. The Best Approximation Theorem

Let W be a subspace of $\mathbb{R}^n$, Let $y \in \mathbb{R}^n$, Let $\hat{y}$ be the orthogonal projection of y onto W. Then $\hat{y}$ is the closest point in W to y, in the sense that

$$\|y - \hat{y}\| < \|y - v\|$$

for all $v \in W$ other than $\hat{y}$.

In example 3.14 we find $\|y - \hat{y}\|$ as follows

This show that the distance between $\mathbf{y}$ and $\hat{\mathbf{y}}$ is $\frac{7}{\sqrt{5}}$. According to the **Best Approximation Theorem** this distance is 'the least' between vectors $\mathbf{y}$ and $\hat{\mathbf{y}}$.

The vector $\hat{\mathbf{y}}$ in theorem 3.6 is still further simplified when $\{\mathbf{u_1}, \mathbf{u_2}, \cdots, \mathbf{u_k}\}$ is an **orthonormal basis of** W. This fact is stated in the next theorem.

Theorem 3.9. *If $\{u_1, u_2, \cdots, u_k\}$ is an orthonormal basis for a subspace W of $\mathbb{R}^n$, then*

$$\hat{y} = \operatorname{proj}_W y = (y \cdot u_1)u_1 + (y \cdot u_2)u_2 + \cdots + (y \cdot u_k)u_k$$

If $U = [u_1\, u_2 \cdots u_k]$ then

$$\hat{y} = \operatorname{proj}_W y = UU^T y \text{ for all } y \in \mathbb{R}^n$$

Proof: The expression of $\operatorname{proj}_W y$ follows immediately from **theorem 3.6** with normalized vector basis. In fact, $\operatorname{proj}_W y$ is a linear combination of columns of matrix U with weight (coefficients) as $y \cdot u_1, y \cdot u_2 \cdots y \cdot u_k$. These weights can be written as $u_1^T y, u_2^T y, \cdots u_k^T y$. These are the entries in $U^T y$. This proves the statement

$$\operatorname{proj}_W y = UU^T y.$$

Hence proved the theorem.

Example 3.15. *Let $W = \operatorname{span}\{u_1, u_2\}$ where*

$$u_1 = \begin{bmatrix} \frac{2}{3} \\ \frac{1}{3} \\ \frac{2}{3} \end{bmatrix}, \quad u_2 = \begin{bmatrix} \frac{-2}{3} \\ \frac{2}{3} \\ \frac{1}{3} \end{bmatrix} \text{ and } y = \begin{bmatrix} 4 \\ 8 \\ 1 \end{bmatrix}$$

Let $U = [u_1\ u_2]$. Compute $U^T U$ and UU^T. Compute $\operatorname{proj}_W y$ and $(UU^T)y$.

Solution: Now, $U = \begin{bmatrix} \frac{2}{3} & \frac{-2}{3} \\ \frac{1}{3} & \frac{2}{3} \\ \frac{2}{3} & \frac{1}{3} \end{bmatrix}$, $U^T = \begin{bmatrix} \frac{2}{3} & \frac{1}{3} & \frac{2}{3} \\ \frac{-2}{3} & \frac{2}{3} & \frac{1}{3} \end{bmatrix}$

$$UU^T = \begin{bmatrix} \frac{8}{9} & \frac{-2}{9} & \frac{2}{9} \\ \frac{-2}{9} & \frac{5}{9} & \frac{4}{9} \\ \frac{2}{9} & \frac{4}{9} & \frac{5}{9} \end{bmatrix}$$

$$U^T U = \begin{bmatrix} 1 & 0 \\ 0 & 1 \end{bmatrix}$$

$\operatorname{proj}_W y = (y \cdot u_1)u_1 + (y \cdot u_2)u_2$ since given basis $\{u_1, u_2\}$ is an orthonormal basis. (easy to verify).

$$y \cdot u_1 = \begin{bmatrix} 4 \\ 8 \\ 1 \end{bmatrix} \cdot \begin{bmatrix} \frac{2}{3} \\ \frac{1}{3} \\ \frac{2}{3} \end{bmatrix} = 4 \times \frac{2}{3} + 8 \times \frac{1}{3} + 1 \times \frac{2}{3} = 6$$

$$y \cdot u_2 = \begin{bmatrix} 4 \\ 8 \\ 1 \end{bmatrix} \cdot \begin{bmatrix} \frac{-2}{3} \\ \frac{2}{3} \\ \frac{1}{3} \end{bmatrix} = 4 \times \frac{-2}{3} + 8 \times \frac{2}{3} + 1 \times \frac{2}{3}$$

$$\frac{-8 + 16 + 1}{\quad} \quad \frac{9}{\quad} = 3$$

$\therefore \text{proj}_W y = 6u_1 + 3u_2$

Lastly,

$$(UU^T)y = \begin{bmatrix} \frac{8}{9} & \frac{-2}{9} & \frac{2}{9} \\ \frac{-2}{9} & \frac{5}{9} & \frac{4}{9} \\ \frac{2}{9} & \frac{4}{9} & \frac{5}{9} \end{bmatrix} \begin{bmatrix} 4 \\ 8 \\ 1 \end{bmatrix}$$

$$= \begin{bmatrix} 2 \\ 4 \\ 5 \end{bmatrix}$$

3.5 Diagonalization of Symmetric Matrices

We have studied the process of diagonalization of a matrix. In this section we apply the process to symmetric matrices. Symmetric matrices are useful in applications.

Definition 3.10. *A matrix A is said to be a symmetric matrix if $A^T = A$.*

We see that a symmetric matrix is necessarily square. Its diagonal entries are arbitrary but non-diagonal elements are present in pairs on opposite side of the main diagonal. Examples of symmetric matrices are,

$$\begin{bmatrix} 1 & 0 \\ 0 & 3 \end{bmatrix}, \begin{bmatrix} 9 & 6 \\ 6 & 1 \end{bmatrix}, \begin{bmatrix} 3 & 1 & 0 \\ 1 & 9 & 6 \\ 0 & 6 & 9 \end{bmatrix}, \begin{bmatrix} 3 & 0 & 0 \\ 0 & 3 & 0 \\ 0 & 0 & 3 \end{bmatrix}$$

Example 3.16. $A = \begin{bmatrix} 1 & 1 \\ 1 & -1 \end{bmatrix}$, $B = \begin{bmatrix} \frac{2}{3} & \frac{2}{3} & \frac{1}{3} \\ 0 & \frac{1}{3} & \frac{-2}{3} \\ \frac{5}{3} & \frac{-4}{3} & \frac{-2}{3} \end{bmatrix}$

Determine which matrices are orthogonal?

Solution: Since $AA^T \neq I_2$ and $BB^T \neq I_3$. The given matrices are not orthogonal.

Example 3.17. *Orthogonally diagonalize the symmetric matrix A given below*

$$A = \begin{bmatrix} 3 & -2 & 4 \\ -2 & 6 & 2 \\ 4 & 2 & 3 \end{bmatrix}$$

Solution: The characteristic equation of A is $-\lambda^3 + 12\lambda^2 - 21\lambda - 98 = 0 = P(\lambda)$ Factorize the cubic equation using synthetic division as

$$\begin{array}{r|rrrr} -2 & -1 & 12 & -21 & -98 \\ & & 2 & -28 & 98 \\ \hline & -1 & 14 & -49 & 0 \end{array}$$

$$\therefore P(\lambda) = (\lambda + 2)(-\lambda^2 + 14\lambda - 49)$$
$$= (\lambda + 2)(-\lambda + 7)^2$$

The eigen values are $\lambda = -2, 7, 7$.

for $\lambda = -2, v_1 = \begin{bmatrix} -1 \\ \frac{-1}{2} \\ 1 \end{bmatrix}$, for $\lambda = 7, \ v_2 = \begin{bmatrix} 1 \\ 0 \\ 1 \end{bmatrix}$, $v_3 = \begin{bmatrix} \frac{-1}{2} \\ 1 \\ 0 \end{bmatrix}$

observe that vectors v_2 and v_3 are linearly independent but not orthogonal. We find projection of v_3 onto v_2 and the component of v_3 orthogonal to v_2 is

$$z_2 = v_3 - \frac{v_3 \cdot v_2}{v_2 \cdot v_2} v_2$$

$$z_2 = \begin{bmatrix} \frac{-1}{2} \\ 1 \\ 0 \end{bmatrix} - \frac{\frac{-1}{2}}{2} \begin{bmatrix} 1 \\ 0 \\ 1 \end{bmatrix} = \begin{bmatrix} \frac{-1}{4} \\ 1 \\ \frac{1}{4} \end{bmatrix}$$

z_2 is a linear combination of v_2 and v_3 and $\{v_2, z_2\}$ is an orthogonal set in the eigen space for $\lambda = 7$. Now, normalize the vectors v_2 and z_2 to obtain orthonormal basis for the eigen space for $\lambda = 7$.

$$\therefore u_1 = \begin{bmatrix} \frac{1}{\sqrt{2}} \\ 0 \\ \frac{1}{\sqrt{2}} \end{bmatrix}, \ u_2 = \begin{bmatrix} \frac{-1}{\sqrt{18}} \\ \frac{4}{\sqrt{18}} \\ \frac{1}{\sqrt{18}} \end{bmatrix}$$

For $\lambda = -2$, the orthonormal basis for the corresponding eigen space is

$$u_3 = \frac{1}{\|v_1\|} v_1 = \frac{1}{\frac{3}{2}} \begin{bmatrix} -1 \\ \frac{-1}{2} \\ 1 \end{bmatrix}$$

$$= \frac{1}{3} \begin{bmatrix} -2 \\ -1 \\ 2 \end{bmatrix} = \begin{bmatrix} \frac{-2}{3} \\ \frac{-1}{3} \\ \frac{2}{3} \end{bmatrix}$$

Hence the set $\{u_1, u_2, u_3\}$ form an orthonormal set.

Let $P = \begin{bmatrix} \frac{1}{\sqrt{2}} & \frac{-1}{\sqrt{18}} & \frac{-2}{3} \\ 0 & \frac{4}{\sqrt{18}} & \frac{-1}{3} \\ \frac{1}{\sqrt{2}} & \frac{1}{\sqrt{18}} & \frac{2}{3} \end{bmatrix} = [u_1 \ u_2 \ u_3]$ and $D = \begin{bmatrix} 7 & 0 & 0 \\ 0 & 7 & 0 \\ 0 & 0 & -2 \end{bmatrix}$

Hence $AP = PD$

Theorem 3.10. *An $n \times n$ matrix A is orthogonally diagonalizable iff A is a symmetric matrix.*

Theorem 3.11. *If A is symmetric then any two eigen vectors from different eigen sapces are orthogonal.*

In the above example 3.17 we have used both these theorms.

The set of eigen values of a matrix A is known as **the spectrum of** A.

Theorem 3.12. The spectral Theorem for symmetric matrices

An $n \times n$ symmetric matrix A has the following properties.

(a) A has n real eigen values, counting multiplicity.

(b) The dimension of the eigen space for each eigen value λ equals the multiplicity of λ as a root of

(c) The eigen spaces are mutually orthogonal, in the sense that eigen vectors corresponding to different eigen values are orthogonal.

(d) A is orthogonally diagonalizable.

The properties mentioned in Theorem 3.11 are very useful in solving problems of symmetric matrices.

Spectral Decomposition of a symmetric matrix A.

Let A be a matrix with orthogonal diagonalization as $A = PDP^{-1}$ where columns of matrix P are orthonormal eigenvectors $u_1, u_2, ..., u_n$ of A that correspond to eigenvalues $\lambda_1, \lambda_2, ..., \lambda_n$. D is a diagonal matrix of n eigenvalues . P is an orthogonal matrix , so $P^{-1} = P^T$.

So,

$$A = P D P^{-1} = P D P^T = \begin{bmatrix} u_1 & u_2 & . & . & u_n \end{bmatrix} \begin{bmatrix} \lambda_1 & 0 & . & . & . & 0 \\ 0 & \lambda_2 & . & . & . & 0 \\ . & & . & . & . & . \\ . & & . & . & . & . \\ 0 & 0 & . & . & . & \lambda_n \end{bmatrix} \begin{bmatrix} u_1^T \\ u_2^T \\ . \\ . \\ u_n^T \end{bmatrix}$$

$$= \begin{bmatrix} \lambda_1 u_1 & \lambda_2 u_2 & . & . & \lambda_n u_n \end{bmatrix} \begin{bmatrix} u_1^T \\ u_2^T \\ . \\ . \\ u_n^T \end{bmatrix}$$

Therefore, $A = \lambda_1 u_2 u_2^T + \lambda_1 u_1 u_1^T + ---+\lambda_n u_n u_n^T$.

This representation of A as a sum of $\lambda_i u_i u_i^T$, $\forall i = 1, 2, .., n$ is called **a spectral decomposition of** A. This is called decomposition because it breakes matrix A into pieces determined by the spectrum(eigenvalues) of A. Each term in the decomposition is an $n \times n$ matrix of rank 1. Every column of $\lambda_i u_i u_i^T$ is a multiple of u_i for every i. Also every matrix $u_i u_i^T$ is a projection matrix in the sense that $\forall x \in \mathbb{R}^n$, the vector $u_i u_i^T x$ is the orthogonal projection of x onto the subspace spanned by $\mathbf{u_i}$.

Example 3.18. *Construct a spectural decomposition of the matrix A that has the orthogonal diagonalization.*

$$A = \begin{bmatrix} 3 & 1 \\ 1 & 3 \end{bmatrix}$$

$$= \begin{bmatrix} \frac{1}{\sqrt{2}} & \frac{-1}{\sqrt{2}} \\ \frac{1}{\sqrt{2}} & \frac{1}{\sqrt{2}} \end{bmatrix} \begin{bmatrix} 4 & 0 \\ 0 & 2 \end{bmatrix} \begin{bmatrix} \frac{1}{\sqrt{2}} & \frac{1}{\sqrt{2}} \\ \frac{-1}{\sqrt{2}} & \frac{1}{\sqrt{2}} \end{bmatrix}$$

Solution: Let $P = [u_1 \ u_2]$ where

$$u_1 = \begin{bmatrix} \frac{1}{\sqrt{2}} \\ \frac{1}{\sqrt{2}} \end{bmatrix}, \quad u_2 = \begin{bmatrix} \frac{-1}{\sqrt{2}} \\ \frac{1}{\sqrt{2}} \end{bmatrix}$$

then by **spectral decompostion** described above we write

$$A = 4u_1\,u_1^T + 2u_2\,u_2^T$$

$$u_1 u_1^T = \begin{bmatrix} \frac{1}{\sqrt{2}} \\ \frac{1}{\sqrt{2}} \end{bmatrix} \begin{bmatrix} \frac{1}{\sqrt{2}} & \frac{1}{\sqrt{2}} \end{bmatrix} = \begin{bmatrix} \frac{1}{2} & \frac{1}{2} \\ \frac{1}{2} & \frac{1}{2} \end{bmatrix}$$

$$u_2 u_2^T = \begin{bmatrix} \frac{-1}{\sqrt{2}} \\ \frac{1}{\sqrt{2}} \end{bmatrix} \begin{bmatrix} \frac{-1}{\sqrt{2}} & \frac{1}{\sqrt{2}} \end{bmatrix} = \begin{bmatrix} \frac{1}{2} & \frac{-1}{2} \\ \frac{-1}{2} & \frac{1}{2} \end{bmatrix}$$

$$\therefore A = 4u_1 u_1^T + 2u_2 u_2^T = \begin{bmatrix} 2 & 2 \\ 2 & 2 \end{bmatrix} + \begin{bmatrix} 1 & -1 \\ -1 & 1 \end{bmatrix}$$

$$= \begin{bmatrix} 3 & 1 \\ 1 & 3 \end{bmatrix}$$

Example 3.19. *Construct a spectral decomposition of the matrix A that has the orthogonal diagonalization.*

$$A = \begin{bmatrix} 1 & 1 & 5 \\ 1 & 5 & 1 \\ 5 & 1 & 1 \end{bmatrix} = \begin{bmatrix} \frac{-1}{\sqrt{2}} & \frac{1}{\sqrt{6}} & \frac{1}{\sqrt{3}} \\ 0 & \frac{-2}{\sqrt{6}} & \frac{1}{\sqrt{3}} \\ \frac{1}{\sqrt{2}} & \frac{1}{\sqrt{6}} & \frac{1}{\sqrt{3}} \end{bmatrix} \begin{bmatrix} -4 & 0 & 0 \\ 0 & 4 & 0 \\ 0 & 0 & 7 \end{bmatrix} \begin{bmatrix} \frac{-1}{\sqrt{2}} & 0 & \frac{1}{\sqrt{2}} \\ \frac{1}{\sqrt{6}} & \frac{-2}{\sqrt{6}} & \frac{1}{\sqrt{6}} \\ \frac{1}{\sqrt{3}} & \frac{1}{\sqrt{3}} & \frac{1}{\sqrt{6}} \end{bmatrix}$$

Solution: Note that eigen values of matrix A are $-4, 4, 7$.

$$\text{Let } u_1 = \begin{bmatrix} \frac{-1}{\sqrt{2}} \\ 0 \\ \frac{1}{\sqrt{2}} \end{bmatrix}, \quad u_2 = \begin{bmatrix} \frac{1}{\sqrt{6}} \\ \frac{-2}{\sqrt{6}} \\ \frac{1}{\sqrt{6}} \end{bmatrix}, \quad u_3 = \begin{bmatrix} \frac{1}{\sqrt{3}} \\ \frac{1}{\sqrt{3}} \\ \frac{1}{\sqrt{3}} \end{bmatrix}$$

Compute, $u_1 u_1^T$, $u_2 u_2^T$ and $u_3 u_3^T$ as follows:

$$u_1 u_1^T = \begin{bmatrix} \frac{1}{2} & 0 & \frac{-1}{2} \\ 0 & 0 & 0 \\ \frac{-1}{2} & 0 & \frac{1}{2} \end{bmatrix}, \quad u_2 u_2^T = \begin{bmatrix} \frac{1}{6} & \frac{-2}{6} & \frac{1}{6} \\ \frac{-2}{\sqrt{6}} & \frac{4}{6} & \frac{-2}{6} \\ \frac{1}{\sqrt{6}} & \frac{-2}{6} & \frac{1}{6} \end{bmatrix}, \quad u_3 u_3^T = \begin{bmatrix} \frac{1}{3} & \frac{1}{3} & \frac{1}{3} \\ \frac{1}{\sqrt{3}} & \frac{1}{3} & \frac{1}{3} \\ \frac{1}{\sqrt{3}} & \frac{1}{3} & \frac{1}{3} \end{bmatrix}$$

By spectral decomposition formula

$$A = (-4)u_1 u_1^T + 4u_2 u_2^T + 7u_3 u_3^T$$

$$= \begin{bmatrix} -2 & 0 & 2 \\ 0 & 0 & 0 \\ 2 & 0 & -2 \end{bmatrix} + \begin{bmatrix} \frac{2}{3} & \frac{-4}{3} & \frac{2}{3} \\ \frac{-4}{3} & \frac{8}{3} & \frac{-4}{3} \\ \frac{2}{3} & \frac{-4}{3} & \frac{2}{3} \end{bmatrix} + \begin{bmatrix} \frac{7}{3} & \frac{7}{3} & \frac{7}{3} \\ \frac{7}{3} & \frac{7}{3} & \frac{7}{3} \\ \frac{7}{3} & \frac{7}{3} & \frac{7}{3} \end{bmatrix}$$

$$= \begin{bmatrix} 1 & 1 & 5 \\ 1 & 5 & 1 \\ 5 & 1 & 1 \end{bmatrix}$$

3.6 Quadratic Forms

Quadratic forms have applications in Design criteria and Optimization. It is used in **signal processing**, differential geometry, economics etc.

Definition 3.11. Quadratic Form

A Quadratic form on $\mathbb{R}^n$ is a function Q on $\mathbb{R}^n$ defined as $Q(x) = x^T A x$ where A is an $n \times n$

Matrix A is called the matrix of the quadratic form.

Example 3.20. *A function $Q : \mathbb{R}^2 \to \mathbb{R}$ defined as $Q(x) = x^T I x$. Here $x \in \mathbb{R}^2$, Let $x = \begin{bmatrix} x_1 \\ x_2 \end{bmatrix}$*

$$x^T I x = [x_1 \; x_2] I \begin{bmatrix} x_1 \\ x_2 \end{bmatrix}$$
$$= [x_1^2 + x_2^2] = \|x\|^2 \quad \text{is a quadratic form.}$$

Example 3.21. *Let Q be defined on $\mathbb{R}^2$. Let $x = \begin{bmatrix} x_1 \\ x_2 \end{bmatrix} \in \mathbb{R}^2$.*

Compute

$$x^T A x = [x_1 \; x_2] \begin{bmatrix} 4 & 0 \\ 0 & 3 \end{bmatrix} \begin{bmatrix} x_1 \\ x_2 \end{bmatrix}$$
$$\therefore Q(x) = 4x_1^2 + 3x_2^2$$

This is the quadratic form. There is no cross product term present in $Q(x)$ as matrix A is a diagonal matrix.

Example 3.22. *If matrix A of the given quadratic form is not a diagonal matrix but a symmetric matrix then there exist cross-product terems as shown in this example.*

Solution: Given

$$A = \begin{bmatrix} 3 & -2 \\ -2 & 7 \end{bmatrix}$$
$$x^T A x = [x_1 \; x_2] \begin{bmatrix} 3 & -2 \\ -2 & 7 \end{bmatrix} \begin{bmatrix} x_1 \\ x_2 \end{bmatrix}$$
$$= [(3x_1 - 2x_2)(-2x_1 + 7x_2)] \begin{bmatrix} x_1 \\ x_2 \end{bmatrix}$$
$$= 3x_1^2 - 2x_1 x_2 - 2x_1 x_2 + 7x_2^2$$
$$= 3x_1^2 - 4x_1 x_2 + 7x_2^2.$$

Example 3.23. *Compute the quadratic form $x^T A x$ where $A = \begin{bmatrix} 5 & \frac{1}{3} \\ \frac{1}{3} & 1 \end{bmatrix}$ and $x = \begin{bmatrix} 1 \\ 3 \end{bmatrix}$.*

Solution:

$$x^T A x = [x_1 \; x_2] A \begin{bmatrix} x_1 \\ x_2 \end{bmatrix}$$
$$= [1 \; 3] \begin{bmatrix} 5 & \frac{1}{3} \\ \frac{1}{3} & 1 \end{bmatrix} \begin{bmatrix} 1 \\ 3 \end{bmatrix}$$
$$= [6 \; \frac{10}{3}] \begin{bmatrix} 1 \\ 3 \end{bmatrix}$$
$$= 16$$

Solution: The coefficients of square forms are on the main diagonal of matrix A.

To make A symmetric the coefficient of $x_i x)j, j \neq i$ must be evently split betweeen $(i, j)^{\text{th}}$ and $(j, i)^{\text{th}}$ entries in A.

$$\therefore A = \begin{bmatrix} 3 & -2 \\ -2 & 5 \end{bmatrix}$$

Example 3.25. *Find matrices of Quadratic forms given below in* $\mathbb{R}^3$

(a) $Q(x) = 3x_1^2 + 2x_2^2 - 5x_3^2 - x_1x_2 + 8x_1x_3 - 4x_2x_3$

(b) $Q(x) = 6x_1x_2 + 4x_1x_3 - 10x_2x_3.$

Solution:

(a) The required matrix has diagonal elements as $3, 2, -5$. The cross coefficients are distributed in the matrix as shown in the previous example.

$$\therefore A = \begin{bmatrix} 3 & -3 & 4 \\ -3 & 2 & -2 \\ 4 & -2 & -5 \end{bmatrix}$$

(b) There are no square terms in the given quadratic form. Hence the diagonal elements will be 0. The rest is clear by applying cross-coefficient rule.

$$\therefore A = \begin{bmatrix} 0 & 3 & 2 \\ 3 & 0 & -5 \\ 2 & -5 & 0 \end{bmatrix}$$

Example 3.26. *Compute the value of quadratic form* $Q(x) = x_1^2 - 8x_1 - 5x_2^2$ *for* $x = \begin{bmatrix} -3 \\ 1 \end{bmatrix}.$

Solution: We have

$$Q(x) = x_1^2 - 8x_1x_2 - 5x_2^2$$
$$Q(-3, 1) = (-3)^2 - 8(-3)(1) - 5(1)^2$$
$$= 9 + 24 - 5$$
$$= 28$$

Many times in calculations and applications of Quadratic forms, the computation becomes easy when $Q(x)$ does not have cross-product terms, that is when the corresponding matrix of $Q(x)$ is a diagonal matrix. But this is not always the case. However, the **change of variable** technique used in Quadratic form eliminates the cross-product terms as and when required. The following example illustrates it in detail.

Example 3.27. *Find the quadratic form without product terms for the matrix* $A = \begin{bmatrix} 1 & -4 \\ -4 & -5 \end{bmatrix}.$

Solution: Let $x \in \mathbb{R}^2$ then the corresponding quadratic form for matrix A is

$$Q(x) = x_1^2 - 8x_1x_2 - 5x_2^2.$$

The product term x_1x_2 is eliminated by using the technique of change of variables as follows.

Step 1: Find eigenvalues of A. Here, $\lambda = 3$ and $\lambda = -7$.

Step 2: Orthogonally diagonalize A.

The associated unit eigenvectors are

$$u_1 = \begin{bmatrix} \frac{2}{\sqrt{5}} \\ \frac{-1}{\sqrt{5}} \end{bmatrix} \quad \text{and} \quad u_2 = \begin{bmatrix} \frac{1}{\sqrt{5}} \\ \frac{2}{\sqrt{5}} \end{bmatrix}$$

Make sure that the vectors u_1, u_2 are orthogonal.

In this case u_1 and u_2 correspond to distinct eigen values hence they are orthogonal.

So $\{u_1, u_2\}$ provides an orthonormal basis of $\mathbb{R}^2$

Step 3: Define matrices P **and** D

$$P = [u_1\ u_2] = \begin{bmatrix} \frac{2}{\sqrt{5}} & \frac{1}{\sqrt{5}} \\ \frac{-1}{\sqrt{5}} & \frac{2}{\sqrt{5}} \end{bmatrix} \quad \text{and} \quad D = \begin{bmatrix} 3 & 0 \\ 0 & -7 \end{bmatrix}$$

Step 4: Change of variables step.

Let $\mathbf{x} = \mathbf{P}\mathbf{y}$ where $\mathbf{x} = \begin{bmatrix} x_1 \\ x_2 \end{bmatrix}$ and $\mathbf{y} = \begin{bmatrix} y_1 \\ y_2 \end{bmatrix}$

Then Compute the Quadratic form with change of variables as follows:

$$\begin{aligned}
x_1^2 - 8x_1x_2 - 5x_2^2 &= x^T A x \\
&= (Py)^T A (Py) \\
&= y^T (P^T A P) y \\
&= y^T D y \qquad \because P^T A P = D \\
&= y^T \begin{bmatrix} 3 & 0 \\ 0 & -7 \end{bmatrix} y \\
&= [y_1\ y_2] \begin{bmatrix} 3 & 0 \\ 0 & -7 \end{bmatrix} \begin{bmatrix} y_1 \\ y_2 \end{bmatrix} \\
&= [3y_1 - 7y_2] \begin{bmatrix} y_1 \\ y_2 \end{bmatrix}
\end{aligned}$$

$$Therefore,\ x_1^2 - 8x_1x_2 - 5x_2^2 = 3y_1^2 - 7y_2^2$$

Step 5: Verify the equality of the quadratic forms calculated in step 4.

Let $x = \begin{bmatrix} 1 \\ -1 \end{bmatrix}$ we find y as follows:

$$y = Py = \begin{bmatrix} \frac{2}{\sqrt{5}} & \frac{-1}{\sqrt{5}} \\ \frac{1}{\sqrt{5}} & \frac{2}{\sqrt{5}} \end{bmatrix} \begin{bmatrix} 1 \\ -1 \end{bmatrix}$$

$$= \begin{bmatrix} \frac{3}{\sqrt{5}} \\ 1 \end{bmatrix}$$

putting $x_1 = 1, x_2 = -1$ and $y_1 = \frac{3}{\sqrt{5}}$, $y_2 = \frac{-1}{\sqrt{5}}$ in the quadratic form in step 4.
we get

$$1^2 - 8 \times 1 \times (-1) - 5 \times (-1)^2 = 4$$

$$3 \times \left(\frac{3}{\sqrt{5}}\right)^2 - 7 \times \left(\frac{-1}{\sqrt{5}}\right)^2 = \frac{20}{5} = 4$$

Hence the two quadratic forms are equal.

The change of variable procces is depicted in the following figure.

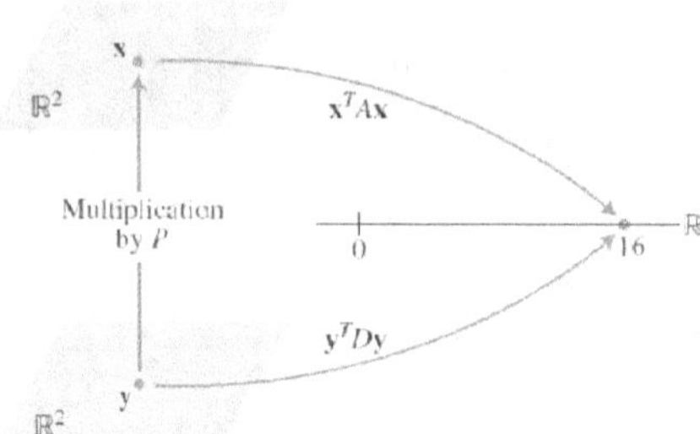

Figure 3.1

The change of variable is illustrated in example 3.27. Now we state the theorem in this context.

Theorem 3.13. *The Principal Axes Theorem*

Let A be an $n \times n$ symmetric matrix. Then there is an orthogonal change of variable, $x = Py$, That transforms the quadratic form $x^T Ax$ into a quadratic form $y^T Dy$ with no cross-product terms.

The columns of matrix P are known as the **Principal axes** of the quadratic from $x^T Ax$. Finding the **principal axes** (determined by the eigen vectors of A) amounts to finding a new coordinate system with respect to which the graph of the function $Q(x) = C, C \in \mathbb{R}$ is in standard position. **If A is a diagonal matrix then the graph is in standard position.** See figures below.

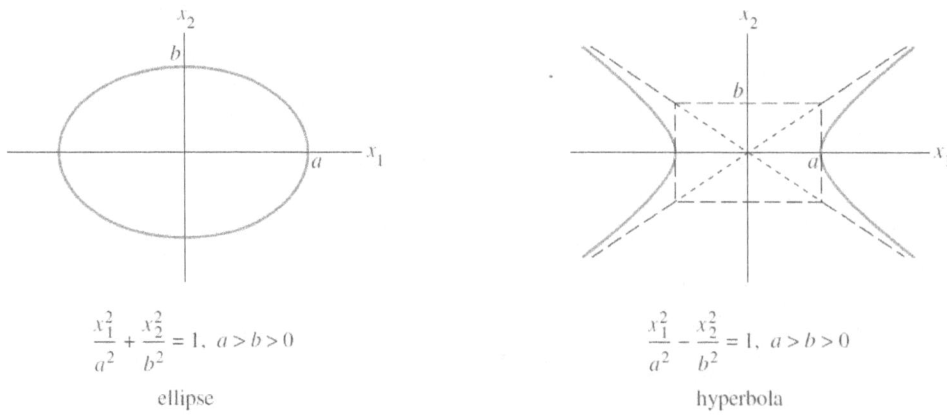

Figure 3.2

If A is **not** a **diagonal matrix** the graph of $Q(x) = C, C \in \mathbb{R}$ is rotated out of standard position see figures below.

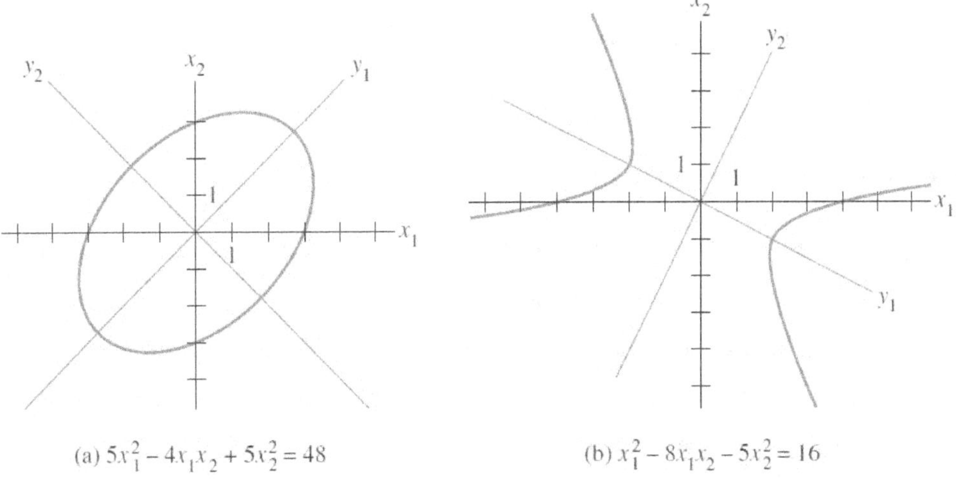

(a) $5x_1^2 - 4x_1x_2 + 5x_2^2 = 48$

(b) $x_1^2 - 8x_1x_2 - 5x_2^2 = 16$

Figure 3.3

Classification of Quadratic forms:

The following are some examples of quadratic forms with domain $\mathbb{R}^2$. The points of $\mathbb{R}^2, (x_1, x_2)$ (say) are mapped on to (x_1, x_2, z). In figure (a) except at $x = 0$ all values of $Q(x)$ are positive. In figure (d) **except at** $x = 0$, all values of $Q(x)$ are negative. The horizontal cross-sections of the graphs are ellipses in figures (a) and (d) as one can imagine. Where as in figure (c) the cross-section is a hyperbola.

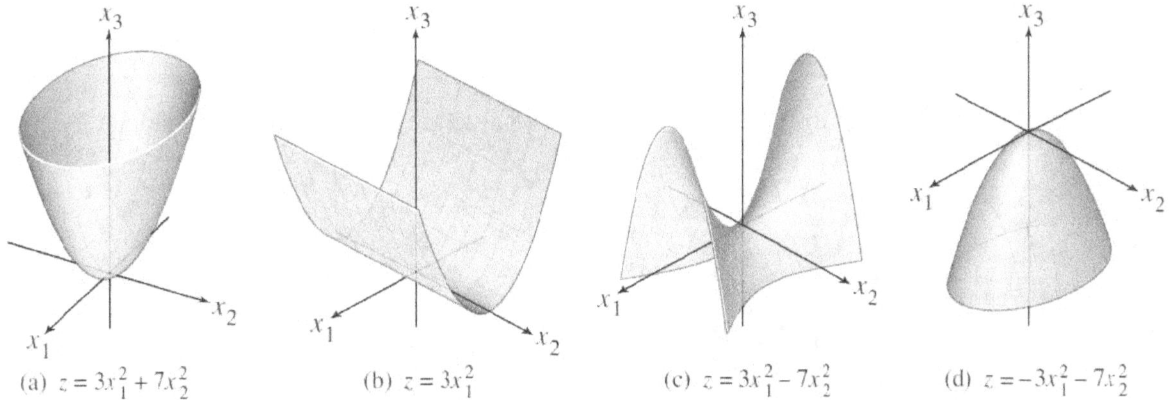

(a) $z = 3x_1^2 + 7x_2^2$

(b) $z = 3x_1^2$

(c) $z = 3x_1^2 - 7x_2^2$

(d) $z = -3x_1^2 - 7x_2^2$

Figure 3.4

Graphs of Quadratic forms.

The following definition classifies the **Quadratic forms** in three types:

Definition 3.12. *A Quadratic form Q is:*

(b) **Negative Definite** *if $Q(x) < 0 \quad \forall x \neq 0$*

(c) **Indefinite** *if $Q(x)$ assumes both positive and negative values.*

In addition to above three types, two more classes of Quadratic forms are defined as:

- **Positive Semidefinite** if $Q(x) \geq 0 \quad \forall x$

- **Negative Semidefinite** if $Q(x) \leq 0 \quad \forall x$

In the above figure, quadratic forms (a) and (b) are both positive semidefinite but (a) is better described as positive definite according to defination.

The next theorem characterizes the quadratic forms in terms of eigen values. this is essential and useful as Geomatrical visualization of the Quadratic form may be challenging in some cases. In such situation one can categorize the Quadratic forms based on their eigen values.

Theorem 3.14. Quadratic Forms and Eigen Values

*Let A be an $n \times n$ **symmetric matrix**. Then the Quadratic form $x^T A x$ is:*

(a) **Positive Definite iff the eigen values of A are all positive**

(b) **Negative Definite iff the eigen values of A are all negative.**

(c) **Indefinite iff A has both negative and positive eigen values**

The proof of the above theorem follows from the fact that, the eigen values $\lambda_1, \lambda_2, \cdots \lambda_n$ controll the sign of $Q(x)$ in the following equation:

$$Q(x) = x^T A x = y^T A y = \lambda_1 y_1^2 + \lambda_2 y_2^2 + \cdots + \lambda_n y_n^2$$

The following figures of Quadratic forms describe the three major types of $Q(x)$ geometrically.

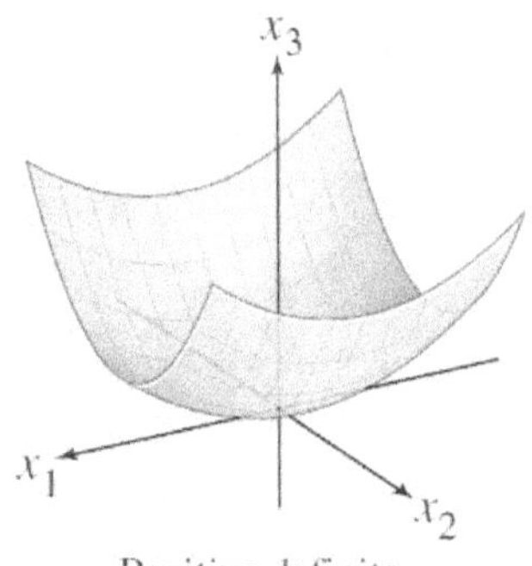

Positive definite

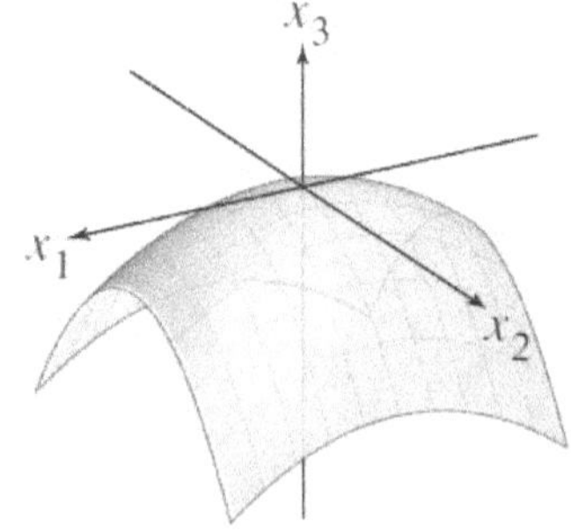

Negative definite

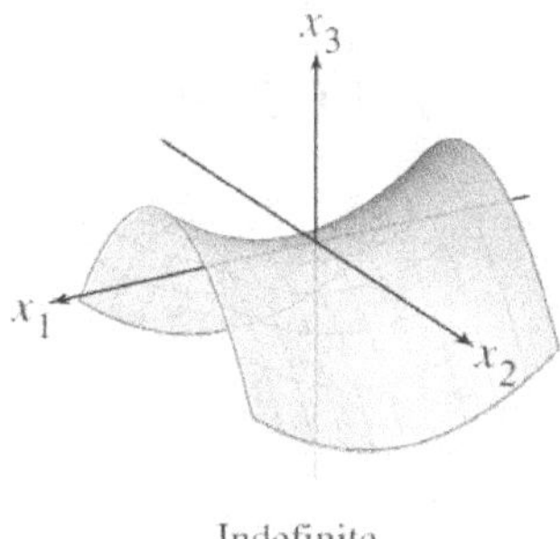

Figure 3.5

Three types of symmetric matrices exist based on the corresponding quadratic forms they characterize as:

A positive definite matrix A if A is symmetric with $Q(x)$ positive.

A negative definite matrix A if A is symmetric with $Q(x)$ negative.

With the same analogy the rest of the types of symmetric matrices are defined.

Example 3.28. *Classify the Quadratic forms given below:*

(a) $4x_1^2 - 4x_1x_2 + 4x_2^2$ (b) $2x_1^2 - 4x_1x_2 - x_2^2$

(c) $x_1^2 - 6x_1x_2 + 9x_2^2$ (d) $-x_1^2 - x_2^2 + 2x_1x_2$

(e) $-x_1^2 - 5x_2^2$

Solution: We use theorem 3.13 to classify the given quadratic form.

(a) $Q(x) = 4x_1^2 - 4x_1x_2 + 4x_2^2$

    eigenvalues are 6 and 2. Hence $Q(x)$ is **positive definite.**

(b) $Q(x) = 2x_1^2 - 4x_1x_2 - x_2^2$

    eigenvalues are 3 and -2. Hence $Q(x)$ is **indefinite.**

(c) $Q(x) = x_1^2 - 6x_1x_2 + 9x_2^2$

    eigenvalues are 10 and 0. Hence $Q(x)$ is **positive semidefinite.**

(d) $Q(x) = -x_1^2 - x_2^2 + 2x_1x_2$

    eigenvalues are 0 **and** -2. Hence $Q(x)$ is **negative semidefinite.**

e) $Q(x) = -x_1^2 - 5x_2^2$

    eigenvalues are $-1, -5$. Hence $Q(x)$ is **negative definite.**

EXERCISE

1. Compute the following quantities using the given vectors:

$$u = \begin{bmatrix} -1 \\ 2 \end{bmatrix}, \quad v = \begin{bmatrix} 4 \\ 6 \end{bmatrix}, \quad w = \begin{bmatrix} 5 \\ 6 \\ -1 \end{bmatrix}, \quad x = \begin{bmatrix} 10 \\ -3 \\ 0 \end{bmatrix}$$

   (a) $u \cdot v, v \cdot v$ (b) $\dfrac{u \cdot v}{v \cdot v}$ (c) $w \cdot w \cdot x$ (d) $\dfrac{w \cdot w}{w \cdot x}$ (e) $\|x\|, \|v\|$ (f) $\left(\dfrac{u \cdot v}{v \cdot v} \right) v.$

2. Find unit vector in the direction of given each vector:

   (a) $\begin{bmatrix} -30 \\ 40 \end{bmatrix}$ (b) $\begin{bmatrix} -6 \\ 4 \\ -3 \end{bmatrix}$ (c) $\begin{bmatrix} 7 \\ -4 \\ 6 \end{bmatrix}$ (d) $\begin{bmatrix} -10 \\ 10 \end{bmatrix}$

3. Compute the distance between given pair of vectors.

   (a) $u = \begin{bmatrix} 2 \\ -5 \\ 1 \end{bmatrix}, \quad v = \begin{bmatrix} -7 \\ -4 \\ 6 \end{bmatrix}$ (b) $x = \begin{bmatrix} \frac{4}{3} \\ -1 \\ \frac{2}{3} \end{bmatrix}, \quad y = \begin{bmatrix} 5 \\ -6 \\ 7 \end{bmatrix}$

4. Determine which pairs of vectors given below are orthogonal.

   (a) $u = \begin{bmatrix} -2 \\ 1 \end{bmatrix}, \quad v = \begin{bmatrix} -3 \\ 1 \end{bmatrix}$ (b) $x = \begin{bmatrix} 3 \\ 2 \\ -5 \\ 0 \end{bmatrix}, \quad y = \begin{bmatrix} -4 \\ 1 \\ -2 \\ 6 \end{bmatrix}$

   (c) $x = \begin{bmatrix} 3 \\ -1 \\ 2 \end{bmatrix}, \quad y = \begin{bmatrix} -1 \\ -1 \\ 1 \end{bmatrix}$ (d) $y = \begin{bmatrix} 1 \\ 1 \\ 1 \end{bmatrix}, \quad v = \begin{bmatrix} -1 \\ 0 \\ 1 \end{bmatrix}$

   (e) $u = \begin{bmatrix} \frac{-2}{3} \\ \frac{1}{3} \\ \frac{2}{3} \end{bmatrix}, \quad v = \begin{bmatrix} \frac{1}{3} \\ \frac{2}{3} \\ 0 \end{bmatrix}.$

5. Determine whether the given sets of vectors are orthogonal or not.

   (a) $\begin{bmatrix} 2 \\ -7 \\ -1 \end{bmatrix}, \begin{bmatrix} -6 \\ -3 \\ 9 \end{bmatrix}, \begin{bmatrix} 3 \\ 1 \\ -1 \end{bmatrix}$ (b) $\begin{bmatrix} 3 \\ -2 \\ 1 \\ 3 \end{bmatrix}, \begin{bmatrix} -1 \\ 3 \\ -3 \\ 4 \end{bmatrix}, \begin{bmatrix} 3 \\ 8 \\ 7 \\ 0 \end{bmatrix}$

   (c) $\begin{bmatrix} 5 \\ -4 \\ 0 \\ 3 \end{bmatrix}, \begin{bmatrix} -4 \\ 1 \\ -3 \\ 8 \end{bmatrix}, \begin{bmatrix} 3 \\ 3 \\ 5 \\ -1 \end{bmatrix}$

6. For the following set of vectors $\{u_1, u_2\}$ or $\{u_1, u_2, y_3\}$ show that each is an orthogonal basis of $\mathbb{R}^2$ or $\mathbb{R}^3$ respectively. Then express vector x as a linear combination of u's.

   (a) $u_1 = \begin{bmatrix} 2 \\ -3 \end{bmatrix}, \quad u_2 = \begin{bmatrix} 6 \\ 4 \end{bmatrix}, \quad x = \begin{bmatrix} 9 \\ -7 \end{bmatrix}$

   (b) $\begin{bmatrix} 3 \end{bmatrix} \quad \begin{bmatrix} -2 \end{bmatrix} \quad \begin{bmatrix} -6 \end{bmatrix}$

(c) $u_1 = \begin{bmatrix} 3 \\ -3 \\ 0 \end{bmatrix}$, $u_2 = \begin{bmatrix} 2 \\ 2 \\ -1 \end{bmatrix}$, $u_3 = \begin{bmatrix} 1 \\ 1 \\ 4 \end{bmatrix}$, $x = \begin{bmatrix} 5 \\ -3 \\ 1 \end{bmatrix}$

7. Compute the orthogonal projection of $y = \begin{bmatrix} 1 \\ 7 \end{bmatrix}$ onto the line through $u = \begin{bmatrix} -4 \\ 2 \end{bmatrix}$ and the origin.

8. For the following vectors y and u write y as the sum of two orthogonal vectors one in Span $\{u\}$ and the other orthogonal to u.

(a) $y = \begin{bmatrix} 2 \\ 3 \end{bmatrix}$, $u = \begin{bmatrix} 4 \\ -7 \end{bmatrix}$ (b) $y = \begin{bmatrix} 2 \\ 6 \end{bmatrix}$, $u = \begin{bmatrix} 7 \\ 1 \end{bmatrix}$

(c) $y = \begin{bmatrix} 3 \\ 1 \end{bmatrix}$, $u = \begin{bmatrix} 8 \\ 6 \end{bmatrix}$ (d) $y = \begin{bmatrix} -3 \\ 9 \end{bmatrix}$, $u = \begin{bmatrix} 1 \\ 2 \end{bmatrix}$

9. Determine which sets of vectors are orthonormal. If a set is only orthogonal normalize the vectors to produce an orthonormal set.

(a) $\begin{bmatrix} 0 \\ 1 \\ 0 \end{bmatrix}$, $\begin{bmatrix} 0 \\ -1 \\ 0 \end{bmatrix}$ (b) $\begin{bmatrix} -0 \cdot 6 \\ 0 \cdot 8 \end{bmatrix}$, $\begin{bmatrix} 0 \cdot 8 \\ 0 \cdot 6 \end{bmatrix}$

(c) $\begin{bmatrix} \frac{-2}{3} \\ \frac{1}{3} \\ \frac{2}{3} \end{bmatrix}$, $\begin{bmatrix} \frac{1}{3} \\ \frac{2}{3} \\ 0 \end{bmatrix}$ (d) $\begin{bmatrix} \frac{1}{\sqrt{18}} \\ \frac{4}{\sqrt{18}} \\ \frac{1}{\sqrt{18}} \end{bmatrix}$, $\begin{bmatrix} \frac{1}{\sqrt{2}} \\ 0 \\ \frac{-1}{\sqrt{2}} \end{bmatrix}$, $\begin{bmatrix} \frac{-2}{3} \\ \frac{1}{3} \\ \frac{-2}{3} \end{bmatrix}$

10. $\{u_1, u_2, u_3, u_4\}$ is an orthogonal basis of $\mathbb{R}^4$. where

$$u_1 = \begin{bmatrix} 0 \\ 1 \\ -4 \\ -1 \end{bmatrix}, \quad u_2 = \begin{bmatrix} 3 \\ 5 \\ 1 \\ 1 \end{bmatrix}, \quad u_3 = \begin{bmatrix} 1 \\ 0 \\ 1 \\ -4 \end{bmatrix}, \quad u_4 = \begin{bmatrix} 5 \\ -3 \\ -1 \\ 1 \end{bmatrix}, \quad x = \begin{bmatrix} 10 \\ -8 \\ 2 \\ 0 \end{bmatrix}$$

Write x as the sum of two vectors, one in Span $\{u_1, u_2, u_3\}$ and the other in Span $\{u_4\}$.

11. $\{u_1, u_2, u_3, u_4\}$ is an orthogonal basis of $W = $ Span $\{u_1, u_2, u_3, u_4\}$.

Write v as a linear combination of u_1, u_2, u_3, u_4. Where

$$u_1 = \begin{bmatrix} 1 \\ 2 \\ 1 \\ 1 \end{bmatrix}, \quad u_2 = \begin{bmatrix} -2 \\ 1 \\ -1 \\ 1 \end{bmatrix}, \quad u_3 = \begin{bmatrix} 1 \\ 1 \\ -2 \\ -1 \end{bmatrix}, \quad u_4 = \begin{bmatrix} -1 \\ 1 \\ 1 \\ -2 \end{bmatrix}, \quad v = \begin{bmatrix} 4 \\ 5 \\ -3 \\ 3 \end{bmatrix}$$

12. Let $y = \begin{bmatrix} 7 \\ 9 \end{bmatrix}$, $u_1 = \begin{bmatrix} \frac{1}{\sqrt{10}} \\ \frac{-3}{\sqrt{10}} \end{bmatrix}$ and $W = $ Span $\{u_1\}$.

(a) Let U be the 2×1 matrix whose only column is u_1. Compute $U^T U$ and $U U^T$.

(b) Compute $\text{proj}_w y$ and $(UU^T)y$.

13. In each of the following verify that the set $\{u_1, u_2\}$ is an orthogonal set, then find the orthogonal

(a) $y = \begin{bmatrix} -1 \\ 4 \\ 3 \end{bmatrix}$, $u_1 = \begin{bmatrix} 1 \\ 1 \\ 0 \end{bmatrix}$, $u_2 = \begin{bmatrix} -1 \\ 1 \\ 0 \end{bmatrix}$

(b) $y = \begin{bmatrix} 6 \\ 3 \\ -2 \end{bmatrix}$, $u_1 = \begin{bmatrix} 3 \\ 4 \\ 0 \end{bmatrix}$, $u_2 = \begin{bmatrix} -4 \\ 3 \\ 0 \end{bmatrix}$

(c) $y = \begin{bmatrix} -1 \\ 2 \\ 6 \end{bmatrix}$, $u_1 = \begin{bmatrix} 3 \\ -1 \\ 2 \end{bmatrix}$, $u_2 = \begin{bmatrix} 1 \\ -1 \\ -2 \end{bmatrix}$

(d) $y = \begin{bmatrix} 6 \\ 4 \\ 1 \end{bmatrix}$, $u_1 = \begin{bmatrix} -4 \\ -1 \\ 1 \end{bmatrix}$, $u_2 = \begin{bmatrix} 0 \\ 1 \\ 1 \end{bmatrix}$

14. Let $W = \text{Span}\ \{u_1, u_2\}$, write vector y as the sum of a vector in W and a vector orthogonal to W.

(a) $y = \begin{bmatrix} 1 \\ 3 \\ 5 \end{bmatrix}$, $u_1 = \begin{bmatrix} 1 \\ 3 \\ -2 \end{bmatrix}$, $u_2 = \begin{bmatrix} 5 \\ 1 \\ 4 \end{bmatrix}$

(b) $y = \begin{bmatrix} -1 \\ 4 \\ 3 \end{bmatrix}$, $u_1 = \begin{bmatrix} 1 \\ 1 \\ 1 \end{bmatrix}$, $u_2 = \begin{bmatrix} -1 \\ 3 \\ -2 \end{bmatrix}$

15. Let $W = \text{Span}\ \{u_1, u_2, u_3\}$, write vector y as the sum of a vector in W and a vector orthogonal to W.

(a) $y = \begin{bmatrix} 4 \\ 3 \\ 3 \\ -1 \end{bmatrix}$, $u_1 = \begin{bmatrix} 1 \\ 1 \\ 0 \\ 1 \end{bmatrix}$, $u_2 = \begin{bmatrix} -1 \\ 3 \\ 1 \\ -2 \end{bmatrix}$, $u_3 = \begin{bmatrix} -1 \\ 0 \\ 1 \\ 1 \end{bmatrix}$

(b) $y = \begin{bmatrix} 3 \\ 4 \\ 5 \\ 6 \end{bmatrix}$, $u_1 = \begin{bmatrix} 1 \\ 1 \\ 0 \\ -1 \end{bmatrix}$, $u_2 = \begin{bmatrix} 1 \\ 0 \\ 1 \\ 1 \end{bmatrix}$, $u_3 = \begin{bmatrix} 0 \\ -1 \\ 1 \\ -1 \end{bmatrix}$

16. Find the closest point to y in the subspace W spanned by v_1 and v_2.

(a) $y = \begin{bmatrix} 3 \\ 1 \\ 5 \\ 1 \end{bmatrix}$, $v_1 = \begin{bmatrix} 3 \\ 1 \\ -1 \\ 1 \end{bmatrix}$, $v_2 = \begin{bmatrix} 1 \\ -1 \\ 1 \\ -1 \end{bmatrix}$

(b) $y = \begin{bmatrix} 3 \\ -1 \\ 1 \\ 13 \end{bmatrix}$, $v_1 = \begin{bmatrix} 1 \\ -2 \\ -1 \\ 2 \end{bmatrix}$, $v_2 = \begin{bmatrix} -4 \\ 1 \\ 0 \\ 3 \end{bmatrix}$

17. Find the best approximation to z by vectors of the form $c_1 v_1 + c_2 v_2$.

(a) $z = \begin{bmatrix} 3 \\ -7 \\ 2 \end{bmatrix}$, $v_1 = \begin{bmatrix} 2 \\ -1 \\ -3 \end{bmatrix}$, $v_2 = \begin{bmatrix} 1 \\ 1 \\ 0 \end{bmatrix}$

(b) $z = \begin{bmatrix} 2 \\ 4 \\ 0 \\ -1 \end{bmatrix}$, $v_1 = \begin{bmatrix} 2 \\ 0 \\ -1 \\ -3 \end{bmatrix}$, $v_2 = \begin{bmatrix} 5 \\ -2 \\ 4 \\ 2 \end{bmatrix}$

18. Find the distance from y to the plane in $\mathbb{R}^3$ spanned by v_1 and v_2 where

$$y = \begin{bmatrix} 5 \\ -9 \\ 5 \end{bmatrix}, \quad v_1 = \begin{bmatrix} -3 \\ -5 \\ 1 \end{bmatrix}, \quad v_2 = \begin{bmatrix} -3 \\ 2 \\ 1 \end{bmatrix}$$

19. Which of the following matrices are symmetric?

(a) $\begin{bmatrix} 3 & 5 \\ 5 & -7 \end{bmatrix}$ (b) $\begin{bmatrix} 3 & -5 \\ -5 & -3 \end{bmatrix}$ (c) $\begin{bmatrix} 2 & 3 \\ 2 & 4 \end{bmatrix}$ (d) $\begin{bmatrix} 0 & 8 & 3 \\ 8 & 0 & -4 \\ 3 & 2 & 0 \end{bmatrix}$

(e) $\begin{bmatrix} -6 & 2 & 0 \\ 2 & -6 & 2 \\ 0 & 2 & -6 \end{bmatrix}$ (f) $\begin{bmatrix} 1 & 2 & 2 & 1 \\ 2 & 2 & 2 & 1 \\ 2 & 2 & 1 & 2 \end{bmatrix}$

20. Determine which of the following matrices are orthogonal?

If orthogonal then find inverse matrix.

(a) $\begin{bmatrix} .6 & .8 \\ .8 & -.6 \end{bmatrix}$ (b) $\begin{bmatrix} \frac{-4}{5} & \frac{3}{5} \\ \frac{3}{5} & \frac{4}{5} \end{bmatrix}$ (c) $\frac{1}{3}\begin{bmatrix} 1 & 2 & 3 \\ 2 & 1 & -2 \\ 2 & -2 & 1 \end{bmatrix}$

(d) $\frac{1}{3}\begin{bmatrix} 2 & 2 & 1 \\ 0 & 1 & -2 \\ 5 & -4 & -2 \end{bmatrix}$ (e) $\frac{1}{2}\begin{bmatrix} 1 & 1 & -1 & -1 \\ 1 & 1 & 1 & 1 \\ 1 & -1 & -1 & 1 \\ 1 & -1 & 1 & -1 \end{bmatrix}$

21. Orthogonally diagonalize the following matrices. Give an orthogonal matrix P and a diagonal

matrix D. For each matrix set of eigen values is provided.

(a) $\begin{bmatrix} 3 & 1 \\ 1 & 3 \end{bmatrix}$ (b) $\begin{bmatrix} 1 & -5 \\ -5 & 1 \end{bmatrix}$ (c) $\begin{bmatrix} 3 & 4 \\ 4 & 9 \end{bmatrix}$ (d) $\begin{bmatrix} 6 & -2 \\ -2 & 9 \end{bmatrix}$

(e) $\begin{bmatrix} 1 & 1 & 5 \\ 1 & 5 & 1 \\ 5 & 1 & 1 \end{bmatrix}$ (f) $\begin{bmatrix} 1 & -6 & 4 \\ -6 & 2 & -2 \\ 4 & -2 & -3 \end{bmatrix}$ (g) $\begin{bmatrix} 3 & -2 & 4 \\ -2 & 6 & 2 \\ 4 & 2 & 3 \end{bmatrix}$ (h) $\begin{bmatrix} 5 & 8 & -4 \\ 8 & 5 & -4 \\ -4 & -4 & -1 \end{bmatrix}$

(i) $\begin{bmatrix} 4 & 3 & 1 & 1 \\ 3 & 4 & 1 & 1 \\ 1 & 1 & 4 & 3 \\ 1 & 1 & 3 & 4 \end{bmatrix}$ (j) $\begin{bmatrix} 4 & 0 & 1 & 0 \\ 0 & 4 & 0 & 1 \\ 1 & 0 & 4 & 0 \\ 0 & 1 & 0 & 4 \end{bmatrix}$ (k) $\begin{bmatrix} 6 & 0 & 0 \\ 0 & 3 & 3 \\ 0 & 3 & 3 \end{bmatrix}$ (l) $\begin{bmatrix} 1 & 1 \\ 1 & 1 \end{bmatrix}$

The corresponding eigen values are as follows:

(e) $-4, 4, 7$ (f) $-3, -6, 9$ (g) $-2, 7$ (h) $-3, 15$ (i) $1, 5, 9$ (j) $3, 5$ (k) $0, 6, 6$ (l) $0, 2$.

22. Construct a spectral decomposition of the matrix A that has the orthogonal decomposition.

(a) $A = \begin{bmatrix} 7 & 2 \\ 2 & 4 \end{bmatrix} = \begin{bmatrix} \frac{2}{\sqrt{5}} & \frac{-1}{\sqrt{5}} \\ \frac{1}{\sqrt{5}} & \frac{2}{\sqrt{5}} \end{bmatrix} \begin{bmatrix} 8 & 0 \\ 0 & 3 \end{bmatrix} \begin{bmatrix} \frac{2}{\sqrt{5}} & \frac{1}{\sqrt{5}} \\ \frac{-1}{\sqrt{5}} & \frac{2}{\sqrt{5}} \end{bmatrix}$

(b) $A = \begin{bmatrix} 3 & 4 \\ 4 & 9 \end{bmatrix} = \begin{bmatrix} \frac{-2}{\sqrt{5}} & \frac{1}{\sqrt{5}} \\ \frac{1}{\sqrt{5}} & \frac{2}{\sqrt{5}} \end{bmatrix} \begin{bmatrix} 1 & 0 \\ 0 & 11 \end{bmatrix} \begin{bmatrix} \frac{-2}{\sqrt{5}} & \frac{1}{\sqrt{5}} \\ \frac{1}{\sqrt{5}} & \frac{2}{\sqrt{5}} \end{bmatrix}$

(c) $A = \begin{bmatrix} 4 & 2 & 2 \\ 2 & 4 & 2 \\ 2 & 2 & 4 \end{bmatrix} = \begin{bmatrix} \frac{-1}{\sqrt{2}} & \frac{-1}{\sqrt{6}} & \frac{1}{\sqrt{3}} \\ \frac{1}{\sqrt{2}} & \frac{-1}{\sqrt{6}} & \frac{1}{\sqrt{3}} \\ 0 & \frac{2}{\sqrt{6}} & \frac{1}{\sqrt{3}} \end{bmatrix} \begin{bmatrix} 2 & 0 & 0 \\ 0 & 2 & 0 \\ 0 & 0 & 8 \end{bmatrix} \begin{bmatrix} \frac{-1}{\sqrt{2}} & \frac{1}{\sqrt{2}} & 0 \\ \frac{-1}{\sqrt{6}} & \frac{-1}{\sqrt{6}} & \frac{2}{\sqrt{6}} \\ \frac{1}{\sqrt{3}} & \frac{1}{\sqrt{3}} & \frac{1}{\sqrt{3}} \end{bmatrix}$

23. State **True** or **False.** Justify each answer.

   a) An $n \times n$ matrix that is orthogonally diagonalizable must be symmetric.

   b) An $n \times n$ symmetric matrix has n distinct real eigen values.

   c) If $A^T = A$ and if vectors u and v satisfy $Au = 3u$ and $Av = 4v$ then $u \cdot v = 0$.

   d) For a non-zero $v \in \mathbb{R}^n$, the matrix vv^T is called a projection matrix.

   e) If $B = PDP^T$, where $P^T = P^{-1}$ and D is a diagonal matrix, then B is a symmetric matrix.

24. Show that if matrix A is symmetric then A^2 is symmetric.

25. Compute the quadratic form $x^T A x$ for $A = \begin{bmatrix} 3 & 2 & 0 \\ 2 & 2 & 1 \\ 0 & 1 & 0 \end{bmatrix}$ and

$$x = \begin{bmatrix} x_1 \\ x_2 \\ x_3 \end{bmatrix}, \quad x = \begin{bmatrix} -2 \\ -1 \\ 5 \end{bmatrix}, \quad x = \begin{bmatrix} \frac{1}{\sqrt{2}} \\ \frac{1}{\sqrt{2}} \\ \frac{1}{\sqrt{2}} \end{bmatrix}$$

26. Find the matrix of quadratic form. Assume $x \in \mathbb{R}^3$.

   (a) $3x_1^2 - 2x_2^2 + 5x_3^2 + 4x_1x_2 - 6x_1x_3$ (b) $4x_3^2 - 2x_1x_2 + 4x_2x_3$.

27. Make a change of variable, $x = Py$ that transforms the following quadratic forms in to quadratic forms with no crossproduct terms. Give matrix P and the new quadratic forms.

   a) $9x_1^2 + 7x_2^2 + 11x_3^2 - 8x_1x_2 + 8x_1x_3$

   b) $x_1^2 + 10x_1x_2 + x_2^2$

   c) $4x_1^2 - 4x_1x_2 + 4x_2^2$

   d) $2x_1^2 + 6x_1x_2 - 6x_2^2$

   e) $2x_1^2 - 4x_1x_2 - x_2^2$

   f) $-x_1^2 - 2x_1x_2 - x_2^2$

   g) $x_1^2 - 6x_1x_2 + 9x_2^2$

   h) $3x_1^2 + 4x_1x_2$

28. Classify the following quadratic forms

a) $2x_1^2 + 5x_2^2 - 6x_1x_2$ b) $7x_1^2 + 5x_1x_2$

c) $x_1^2 - 3x_2^2 + 5x_3^2$ d) $x_1^2 + 2x_2^2 + 2x_1x_2$

e) $x_1^2 - 2x_1x_2$ f) $x_1^2 - x_2^2 + x_3^2 - 2x_1x_2 + 4x_2x_3$

g) $5x_1^2 + 2x_2^2 - x_1x_2$ h) $2x_1^2 + x_2^2 + 3x_1x_2$

i) $x_1^2 + 2x_1x_2 + x_2^2$ j) $-x_1^2 - 3x_2^2$

k) $2x_1x_2 + 2x_2x_3 + 2x_1x_3$ l) $-x_1^2 - x_2^2 + 2x_1x_2.$

SOLUTIONS

1. a) $8,52$ b) $\frac{8}{52}$ c) $\begin{bmatrix} 620 \\ -186 \\ 0 \end{bmatrix}$ d) $\frac{62}{32}$ e) $\sqrt{109}.\sqrt{52}$ f) $\begin{bmatrix} \frac{32}{52} \\ \frac{48}{52} \end{bmatrix}$

2. a) $\begin{bmatrix} \frac{-3}{5} \\ \frac{4}{5} \end{bmatrix}$ b) $\begin{bmatrix} \frac{-6}{\sqrt{61}} \\ \frac{4}{\sqrt{61}} \\ \frac{-3}{\sqrt{61}} \end{bmatrix}$ c) $\begin{bmatrix} \frac{7}{\sqrt{101}} \\ \frac{-4}{\sqrt{101}} \\ \frac{6}{\sqrt{101}} \end{bmatrix}$ d) $\begin{bmatrix} \frac{-1}{\sqrt{2}} \\ \frac{1}{\sqrt{2}} \end{bmatrix}$

3. a) $\sqrt{107}$ b) $\frac{\sqrt{707}}{3}$

4. a) Not orthogonal b) Yes orthogonal c) Yes orthogonal d) Yes orthogonal e) Yes orthogonal

5. a) Not orthogonal b) Yes orthogonal c) Not orthogonal

6.   a) $x = 3.u_1 + \frac{1}{2}.u_2$

    b) $x = \frac{-3}{2}.u_1 + \frac{3}{4}.u_2$

    c) $x = \frac{4}{3}.u_1 + \frac{1}{3}.u_2 + \frac{1}{3}.u_3$

7. $\begin{bmatrix} -2 \\ 1 \end{bmatrix}$

8.   a) $y = \begin{bmatrix} \frac{-4}{5} \\ \frac{7}{5} \end{bmatrix} + \begin{bmatrix} \frac{14}{5} \\ \frac{8}{5} \end{bmatrix}$

    b) $y = \begin{bmatrix} \frac{14}{5} \\ \frac{2}{5} \end{bmatrix} + \begin{bmatrix} \frac{-4}{5} \\ \frac{28}{5} \end{bmatrix}$

    c) $y = \begin{bmatrix} \frac{12}{5} \\ \frac{9}{5} \end{bmatrix} + \begin{bmatrix} \frac{3}{5} \\ \frac{-4}{5} \end{bmatrix}$

    d) $y = \begin{bmatrix} -3 \\ -6 \end{bmatrix} + \begin{bmatrix} 0 \\ 15 \end{bmatrix}$

9.   a) Not orthogonal hence not orthonormal

    b) orthonormal

    c) orthogonal but not orthonrmal. Orthonormal set= $\left\{ \begin{bmatrix} \frac{-2}{3} \\ \frac{1}{3} \\ \frac{2}{3} \end{bmatrix} , \begin{bmatrix} \frac{1}{\sqrt{5}} \\ \frac{2}{\sqrt{5}} \\ 0 \end{bmatrix} \right\}$

    d) Orthonormal

10. $x = \frac{-8}{9}u_1 - \frac{2}{9}u_2 + \frac{2}{3}u_3 + 2u_4$

$$x = \begin{bmatrix} 0 \\ -2 \\ 4 \\ -2 \end{bmatrix} + \begin{bmatrix} 10 \\ -6 \\ -2 \\ 2 \end{bmatrix}$$

11. $v = \frac{14}{7}u_1 + \frac{3}{7}u_2 + \frac{12}{7}u_3 - \frac{8}{7}u_4$

12. a) $U^T U = [1], \quad UU^T = \frac{1}{10}\begin{bmatrix} 1 & -3 \\ -3 & 9 \end{bmatrix}$

   b) $proj_W y = \frac{-20}{\sqrt{10}}u_1, \quad (UU^T)y = \begin{bmatrix} -2 \\ 6 \end{bmatrix}$

13. a) $\begin{bmatrix} -1 \\ 4 \\ 0 \end{bmatrix}$ b) $\begin{bmatrix} 6 \\ 3 \\ 0 \end{bmatrix}$ c) $\begin{bmatrix} -1 \\ 2 \\ 6 \end{bmatrix}$ d) $\begin{bmatrix} 6 \\ 4 \\ 1 \end{bmatrix}$

14. a) $y = \begin{bmatrix} \frac{10}{3} \\ \frac{32}{3} \\ \frac{8}{3} \end{bmatrix} + \begin{bmatrix} \frac{-7}{3} \\ \frac{7}{3} \\ \frac{7}{3} \end{bmatrix}$

   b) $y = \begin{bmatrix} \frac{3}{2} \\ \frac{7}{2} \\ 1 \end{bmatrix} + \begin{bmatrix} \frac{-5}{2} \\ \frac{1}{2} \\ 2 \end{bmatrix}$

15. a) $y = \begin{bmatrix} 2 \\ 4 \\ 0 \\ 0 \end{bmatrix} + \begin{bmatrix} 2 \\ -1 \\ 3 \\ -1 \end{bmatrix}$ b) $y = \begin{bmatrix} 5 \\ 2 \\ 3 \\ 6 \end{bmatrix} + \begin{bmatrix} -2 \\ 2 \\ 2 \\ 0 \end{bmatrix}$

16. a) $\begin{bmatrix} 3 \\ -1 \\ 1 \\ -1 \end{bmatrix}$ b) $\begin{bmatrix} 4 \\ 4 \\ 2 \\ 8 \end{bmatrix}$

17. a) $\begin{bmatrix} -1 \\ -3 \\ -2 \\ 3 \end{bmatrix}$ b) $\begin{bmatrix} 1 \\ 0 \\ \frac{-1}{2} \\ \frac{-3}{2} \end{bmatrix}$

18. $\sqrt{40}$

19. symmetric matrices :a),b),e)

20. orthogonal matrices :a),b),e)

21. a) $P = \begin{bmatrix} \frac{1}{\sqrt{2}} & \frac{-1}{\sqrt{2}} \\ \frac{1}{\sqrt{2}} & \frac{1}{\sqrt{2}} \end{bmatrix}, \quad D = \begin{bmatrix} 4 & 0 \\ 0 & 2 \end{bmatrix}$

   b) $P = \begin{bmatrix} -\frac{1}{\sqrt{2}} & \frac{-1}{\sqrt{2}} \\ \frac{-1}{\sqrt{2}} & \frac{1}{\sqrt{2}} \end{bmatrix}, \quad D = \begin{bmatrix} -4 & 0 \\ 0 & 6 \end{bmatrix}$

   c) $P = \begin{bmatrix} \frac{-2}{\sqrt{5}} & \frac{1}{\sqrt{5}} \\ & \end{bmatrix} \quad D = \begin{bmatrix} 1 & 0 \\ & \end{bmatrix}$

d) $P = \begin{bmatrix} \frac{-1}{2} & 2 \\ 1 & 1 \end{bmatrix}$, $D = \begin{bmatrix} 10 & 0 \\ 0 & 5 \end{bmatrix}$

e) $P = \begin{bmatrix} \frac{-1}{\sqrt{2}} & \frac{1}{\sqrt{6}} & \frac{1}{\sqrt{3}} \\ 0 & \frac{-2}{\sqrt{6}} & \frac{1}{\sqrt{3}} \\ \frac{1}{\sqrt{2}} & \frac{1}{\sqrt{6}} & \frac{1}{\sqrt{3}} \end{bmatrix}$, $D = \begin{bmatrix} -4 & 0 & 0 \\ 0 & 4 & 0 \\ 0 & 0 & 7 \end{bmatrix}$

f) $P = \begin{bmatrix} 0.6 & \frac{1}{3} & 0.6 \\ \frac{1}{3} & 0.6 & -0.6 \\ -0.6 & 0.6 & \frac{1}{3} \end{bmatrix}$, $D = \begin{bmatrix} -6 & 0 & 0 \\ 0 & -3 & 0 \\ 0 & 0 & 9 \end{bmatrix}$

g) $P = \begin{bmatrix} \frac{-1}{\sqrt{5}} & \frac{4}{\sqrt{45}} & \frac{-2}{3} \\ \frac{2}{\sqrt{5}} & \frac{2}{\sqrt{45}} & \frac{-1}{3} \\ 0 & \frac{5}{\sqrt{45}} & \frac{2}{3} \end{bmatrix}$, $D = \begin{bmatrix} 7 & 0 & 0 \\ 0 & 7 & 0 \\ 0 & 0 & -2 \end{bmatrix}$

h) $P = \begin{bmatrix} 0.29 & 0.179 & 0.939 \\ 0.686 & 0.723 & -0.073 \\ 0.66 & 0.66 & -0.33 \end{bmatrix}$, $D = \begin{bmatrix} -3 & 0 & 0 \\ 0 & -3 & 0 \\ 0 & 0 & 15 \end{bmatrix}$

i) $P = \begin{bmatrix} 0 & \frac{1}{\sqrt{2}} & \frac{1}{2} & \frac{1}{2} \\ 0 & -\frac{1}{\sqrt{2}} & \frac{1}{2} & \frac{1}{2} \\ \frac{1}{\sqrt{2}} & 0 & -\frac{1}{2} & \frac{1}{2} \\ -\frac{1}{\sqrt{2}} & 0 & \frac{1}{2} & \frac{1}{2} \end{bmatrix}$, $D = \begin{bmatrix} 1 & 0 & 0 & 0 \\ 0 & 1 & 0 & 0 \\ 0 & 0 & 5 & 0 \\ 0 & 0 & 0 & 9 \end{bmatrix}$

j) $P = \begin{bmatrix} -\frac{1}{\sqrt{2}} & 0 & 0 & -\frac{1}{\sqrt{2}} \\ 0 & \frac{1}{\sqrt{2}} & \frac{1}{\sqrt{2}} & 0 \\ \frac{1}{\sqrt{2}} & 0 & 0 & -\frac{1}{\sqrt{2}} \\ 0 & -\frac{1}{\sqrt{2}} & \frac{1}{\sqrt{2}} & 0 \end{bmatrix}$, $D = \begin{bmatrix} 3 & 0 & 0 & 0 \\ 0 & 3 & 0 & 0 \\ 070 & 5 & 0 \\ 0 & 0 & 0 & 5 \end{bmatrix}$

k) $P = \begin{bmatrix} 0 & 0 & 1 \\ -\frac{1}{\sqrt{2}} & \frac{1}{\sqrt{2}} & 0 \\ \frac{1}{\sqrt{2}} & \frac{1}{\sqrt{2}} & 0 \end{bmatrix}$, $D = \begin{bmatrix} 0 & 0 & 0 \\ 0 & 6 & 0 \\ 0 & 0 & 6 \end{bmatrix}$

l) $P = \begin{bmatrix} -\frac{1}{\sqrt{2}} & \frac{1}{\sqrt{2}} \\ \frac{1}{\sqrt{2}} & \frac{1}{\sqrt{2}} \end{bmatrix}$, $D = \begin{bmatrix} 0 & 0 \\ 0 & 2 \end{bmatrix}$

22. a) $A = 8u_1 u_1^T + 3u_2 u_2^T$

    b) $A = u_1 u_1^T + 11u_2 u_2^T$

    c) $A = 2u_1 u_1^T + 2u_2 u_2^T + 8u_3 u_3^T$

23. a)True b)False c)True d)True e)True

25. $Q(x) = 3x_1{}^2 + 4x_1 x_2 + x_2 x_3 + 2x_2{}^2$

    $Q(\begin{bmatrix} -2 \\ -1 \\ 5 \end{bmatrix}) = 17$, $Q(\begin{bmatrix} \frac{1}{\sqrt{2}} \\ \frac{1}{\sqrt{2}} \\ \frac{1}{\sqrt{2}} \end{bmatrix}) = 5$

26 a) $\begin{bmatrix} 3 & 2 & -3 \\ 2 & -2 & 0 \\ -3 & 0 & 5 \end{bmatrix}$ b) $\begin{bmatrix} 0 & -1 & 0 \\ -1 & 0 & 2 \\ 0 & 2 & 4 \end{bmatrix}$

27. a) $P = \begin{bmatrix} 0.66 & -\frac{1}{3} & 0.66 \\ 0 66 & 0 66 & -\frac{1}{} \end{bmatrix}$ $Q(y) = 3y_1{}^2 + 9y_2{}^2 + 15y_3{}^2$

b) $P = \begin{bmatrix} -\frac{1}{\sqrt{2}} & \frac{1}{\sqrt{2}} \\ \frac{1}{\sqrt{2}} & \frac{1}{\sqrt{2}} \end{bmatrix}$, $Q(y) = -4y_1{}^2 + 6y_2{}^2$

c) $P = \begin{bmatrix} -\frac{1}{\sqrt{2}} & -\frac{1}{\sqrt{2}} \\ -\frac{1}{\sqrt{2}} & \frac{1}{\sqrt{2}} \end{bmatrix}$, $Q(y) = 2y_1{}^2 + 6y_2{}^2$

d) $P = \begin{bmatrix} -\frac{1}{\sqrt{10}} & \frac{-3}{\sqrt{10}} \\ \frac{3}{\sqrt{10}} & -\frac{1}{\sqrt{10}} \end{bmatrix}$, $Q(y) = -7y_1{}^2 + 3y_2{}^2$

e) $P = \begin{bmatrix} -\frac{1}{\sqrt{5}} & -\frac{2}{\sqrt{5}} \\ -\frac{2}{\sqrt{5}} & \frac{1}{\sqrt{5}} \end{bmatrix}$, $Q(y) = -2y_1{}^2 + 3y_2{}^2$

f) $P = \begin{bmatrix} \frac{1}{\sqrt{2}} & -\frac{1}{\sqrt{2}} \\ \frac{1}{\sqrt{2}} & \frac{1}{\sqrt{2}} \end{bmatrix}$, $Q(y) = -2y_1{}^2$

g) $P = \begin{bmatrix} -\frac{3}{\sqrt{10}} & -\frac{1}{\sqrt{10}} \\ \frac{1}{\sqrt{10}} & \frac{3}{\sqrt{10}} \end{bmatrix}$, $Q(y) = 10y_2{}^2$

h) $P = \begin{bmatrix} \frac{1}{\sqrt{5}} & -\frac{2}{\sqrt{5}} \\ -\frac{2}{\sqrt{5}} & -\frac{1}{\sqrt{5}} \end{bmatrix}$, $Q(y) = -y_1{}^2 + 4y_2{}^2$

28. a) positive definite

b) indefinite

c) indefinite

d) positive definite

e) positive semidefinite

f) indefinite

g) positive definite

h) indefinite

i) positive semidefinite

j) negative definite

k) indefinite

l) negative semidefinite

◆◆◆

Chapter 4

The Geometry of Vector Spaces

4.1   Introduction

We have studied finite dimensional vector spaces for centuries the mathematicians are studying the geometry of objects in three dimensions. But now people from mathematical fraternity deal with objects in vector spaces with higher dimensions like four, five or even hundred regularly. Such study of geometry of higher dimensions not only provides new ways of visualizing abstract algebraic concepts, but also creates tools that may be applied in $\mathbb{R}^3$. These tools are of immense interest in computer graphics.

This chapter describes and focusses on the set of vectors that can be visualized as geometric objects such as line segments, polygons and solid objects. Individual vectors are viewed as points. The sets of vectors are described by linear combinations with various restrictions on weight (coefficients). The visualization is mainly in $\mathbb{R}^2$ and $\mathbb{R}^3$, however the concept can be extented to $\mathbb{R}^n$ and other vector spaces.

4.2   Affine Combinations

Definition 4.1. Affine combination of vectors:

Given vectors (or "points"). $\mathbf{v_1}, \mathbf{v_2}, \cdots \mathbf{v_k}$ *in* $\mathbb{R}^n$ *and scalars* $\mathbf{c_1}, \mathbf{c_2}, \cdots \mathbf{c_k}$ *then a linear combination of vectors.*

$$c_1 \, v_1 + c_2 \, v_2 + \cdots + c_k \, v_k$$

Such that $c_1 + c_2 + \cdots + c_k = 1,$ *is called* **an affine combination** *of* $v_1 \, v_2, \cdots, v_k$

For example: Let $v_1 v_2, v_3 \in \mathbb{R}^3$, $c_1 = 2$, $c_2 = -1$, $c_3 = 0$ then $c_1 \, v_1 + c_2 \, v_2 + c_3 \, v_3 = 2v_1 - v_2 + 0v_3$ is an affine combination of the vectors v_1, v_2 and v_3 in $\mathbb{R}^3$, as $c_1 + c_2 + c_3 = 1$.

Definition 4.2. Affine hull (or span)

The set of all affine combinations of vectors (points) in a set **S** *is called the affine hull (or affine span) of* **S***. It is denoted by* aff **S***.*

For example: (a) Affine hull of a single vector v_1 is $\{\mathbf{v_1}\}$ as $c_1 = 1$ is the only choice of c_1.

(b) For two distinct vectors $\mathbf{v_1}, \mathbf{v_2}$ the special way of writing $\mathbf{c_1}$ and $\mathbf{c_2}$ is as follows:

$\mathbf{c_1 = t}$ and $\mathbf{c_2 = 1 - t}$. So that $c_1 + c_2 = 1 \; \forall \; t \in \mathbb{R}$. **aff S** = set of all linear combinations of v_1, v_2 such that $c_1 + c_2 = 1$. That is, for $S = \{v_1, v_2\}$

Let $y \in$ aff S then

$$\mathbf{y = tv_1 + (1 - t)v_2, \; t \in \mathbb{R}}$$

. The set aff S is non-empty, as $v_1, v_2 \in$ aff S (for $t = 1$ and $t = 0$ respectively). Geometrically, whenever $v_1 \neq v_2$, **aff S describes a line through** v_1 **and** v_2. (See figure below.)

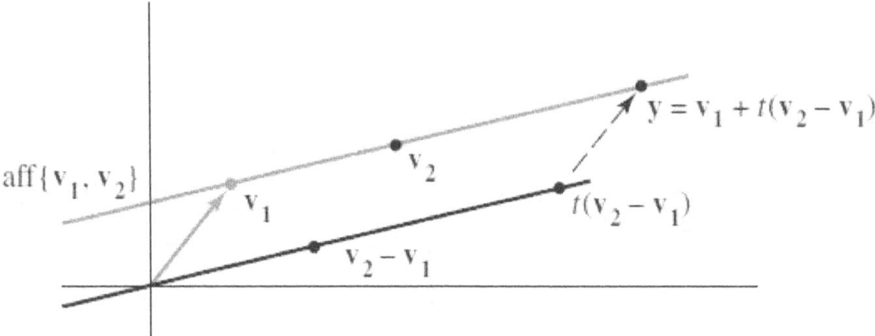

Figure 4.1

from figure we see that point y,

$$\mathbf{y = v_1 + t(v_2 - v_1)}$$

is an affine combination of $\mathbf{v_1}$ and $\mathbf{v_2}$. Note that, $(\mathbf{y - v_1}) = \mathbf{t(v_2 - v_1)}$. That is the point (vector) $\mathbf{y - v_1}$ is a multiple of point (vector) $\mathbf{v_2 - v_1}$. This point $y - v_1$ is known as a translated point (or vector). The following theorem describes the relation between $\mathbf{y}$ and $\mathbf{y - v_1}$.

Theorem 4.1. *A point* $\mathbf{y}$ *in* $\mathbb{R}^n$ *is as affine combination* $\mathbf{v_1, v_2, \cdots v_p}$ *in* $\mathbb{R}^n$ *iff* $\mathbf{y - v_1}$ *is a linear combinationof the translated points* $v_2 - v_1, v_3 - v_1, \cdots v_p - v_1$.

Proof: Suppose $\mathbf{y - v_1}$ is a linear combination of $v_2 - v_1, v_3 - v_1, \cdots v_p - v_1$. That is,

$$y - v_1 = c_2(v_2 - v_1) + c_3(v_3 - v_1) \cdots + c_p(v_p - v_1)$$

$$\therefore y = (1 - c_2 - c_3 \cdots c_p)v_1 + c_2 v_2 + c_3 v_3 + \cdots + c_p v_p$$

Hence the weights of the linear combination $(1 - c_2 - c_3 \cdots - c_p), c_2, c_3 \cdots c_p$ add up to 1.

It proves that y is an affine combination of $v_1, v_2 \cdots v_p$.

Conversely, Suppose points $v_1, v_2 \cdots v_p$ are such that

$$y = c_1 v_1 + c_2 v_2 + \cdots + c_p v_p$$

with $c_1 + c_2 + \cdots + c_p = 1$.

That is $c_1 = 1 - c_2 - c_3 - \cdots - c_p$. Substituting c_1,

So,

$$y = (1 - c_2 - c_3 - \cdots c_p)v_1 + c_2 v_2 + \cdots + c_p v_p$$

$$\therefore y - v_1 = c_2(v_2 - v_1) + c_3(v_3 - v_1) + \cdots + c_p(v_p - v_1)$$

This shows that $\mathbf{y} - \mathbf{v_1}$ is a linear combination of $v_2 - v_1, v_3 - v_1, \cdots v_p - v_1$.

Hence, proved the theorem.

Theorem (4.1) is true for all v_i whenever v_1 is replaced by v_i in the statement.

Example 4.1. *If possible, write point* $\mathbf{y}$ *as an affine combination of* $\mathbf{v_1}, \mathbf{v_2}, \mathbf{v_3}, \mathbf{v_4}$.
where $v_1 = \begin{bmatrix} 1 \\ 2 \end{bmatrix}$, $v_2 = \begin{bmatrix} 2 \\ 5 \end{bmatrix}$, $v_3 = \begin{bmatrix} 1 \\ 3 \end{bmatrix}$, $v_4 = \begin{bmatrix} -2 \\ 2 \end{bmatrix}$, *and* $y = \begin{bmatrix} 4 \\ 1 \end{bmatrix}$.

Solution: Compute the vectors $v_2 - v_1, v_3 - v_1, v_4 - v_1, y - v_1$ as
$$v_2 - v_1 = \begin{bmatrix} 1 \\ 3 \end{bmatrix}, \ v_3 - v_1 = \begin{bmatrix} 0 \\ 1 \end{bmatrix}, \ v_4 - v_1 = \begin{bmatrix} -3 \\ 0 \end{bmatrix}, \ and \ y - v_1 = \begin{bmatrix} 3 \\ -1 \end{bmatrix}$$
To find the scalars c_2, c_3, c_4 such that $c_2 + c_3 + c_4 = 1$.

We simplify the augmented matrix corresponding to the equation.

$$c_2(v_2 - v_1) + c_3(v_3 - v_1) + c_4(v_4 - v_1) = y - v_1$$

$$\begin{bmatrix} 1 & 0 & -3 & 3 \\ 3 & 1 & 0 & -1 \end{bmatrix} \sim \begin{bmatrix} 1 & 0 & -3 & 3 \\ 0 & 1 & 9 & -10 \end{bmatrix}$$

Observe that, two leading elements exist in the echelon form. c_4 is a free variable. The general solution is

$$c_2 = 3c_4 + 3$$

$$c_3 = -9c_4 - 10$$

$$c_4 \qquad \text{is free}$$

In particular, Let $c_4 = 0$

then $c_2 = 3$, $c_3 = -10$, $c_4 = 0$

Substituting the values we get

$$y - v_1 = 3(v_2 - v_1) - 10(v_3 - v_1) + 0(v_4 - v_1)$$

$$\therefore y = 8v_1 + 3v_2 - 10v_3$$

Observe that in the above linear combination the sum of the coefficients is $1\,(8 + 3 - 10)$ hence it is the required affine combination of vectors $\mathbf{v_1}, \mathbf{v_2}, \mathbf{v_3}, \mathbf{v_4}$ that expresses $\mathbf{y}$.

The method explained in example 4.3 works for any arbitrary set of vectors $\{v_1, v_2, \cdots v_k\}$. In parto the $\mathbb{B}$-**coordinates** of vector $\mathbf{y}$. whenever $\mathbb{B} = \{\mathbf{b_1}, \mathbf{b_2} \cdots \mathbf{b_n}\}$ is a basis of $\mathbb{R}^n$. That is, let $\mathbb{B} = \{\mathbf{b_1}, \mathbf{b_2} \cdots \mathbf{b_n}\}$ be a basis of $\mathbb{R}^n$, then any $\mathbf{y} \in \mathbb{R}^\mathbf{n}$ is a unique linear combination of $b_1, b_2 \cdots b_n$. **This combination is an affine combination of b's iff the weights sum is 1. (These weight are the B-coordinates of** y**)**.

The following example illustrates it.

Example 4.2. *Let* $\mathbb{B} = \{\mathbf{b_1}, \mathbf{b_2}, \mathbf{b_3}\}$ *be a basis of* $\mathbb{R}^3$*, and* $\mathbf{P_1}, \mathbf{P_2}$ *be two vectors of* $\mathbb{R}^3$*. Determine whether* $\mathbf{P_1}$ *and* $\mathbf{P_2}$ *are affine combinations of vectors of* $\mathbb{B}$*.*

Solution: Let $b_1 = \begin{bmatrix} 4 \\ 0 \\ 3 \end{bmatrix}$, $b_2 = \begin{bmatrix} 0 \\ 4 \\ 2 \end{bmatrix}$ $b_3 = \begin{bmatrix} 5 \\ 2 \\ 4 \end{bmatrix}$ $P_1 = \begin{bmatrix} 2 \\ 0 \\ 0 \end{bmatrix}$ $P_2 = \begin{bmatrix} 1 \\ 2 \\ 2 \end{bmatrix}$. As per above discussion we wish to find the $\mathbb{B}$-coordinates of vectors $\mathbf{P_1}$ and $\mathbf{P_2}$. This can be done confinedly as follows:

$$\begin{bmatrix} 4 & 0 & 5 & 2 & 1 \\ 0 & 4 & 2 & 0 & 2 \\ 3 & 2 & 4 & 0 & 2 \end{bmatrix} \sim \begin{bmatrix} 1 & 0 & 0 & -2 & \frac{2}{3} \\ 0 & 1 & 0 & -1 & \frac{2}{3} \\ 0 & 0 & 1 & 2 & \frac{-1}{3} \end{bmatrix}$$

column 4 of the reduced matrix corresponds to build $\mathbf{P_1}$ as $\mathbf{P_1} = -2\mathbf{b_1} - \mathbf{b_2} + 2\mathbf{b_3}$, Here the sum of the $\mathbb{B}$-coordinates (coefficients of b_1, b_2, b_3) of $\mathbf{P_1}$ is $-1 \, (\neq 1)$. Hence $\mathbf{P_1}$ is **not** an affine comabination of b_1, b_2, b_3.

Column 5 of the reduced matrix corresponds to build $\mathbf{P_2}$ as $\mathbf{P_2} = \frac{2}{3}\mathbf{b_1} + \frac{2}{3}\mathbf{b_2} - \frac{1}{3}\mathbf{b_3}$. Here the sum of the $\mathbb{B}$-coordinates (coefficients of b_1, b_2, b_3) of $\mathbf{P_2}$ is 1. Hence $\mathbf{P_2}$ is an affine combination of b_1, b_2, b_3.

Definition 4.3. An affine set

A set $\mathbf{S}$ *is affine if* $p, q \in S$ *implies that* $(1 - t)p + tq \in S$ *for each real number* t*.*

The following theorem characterizes the affine set.

Theorem 4.2. *A set* $\mathbf{S}$ *is affine iff every affine combination of points of* $\mathbf{S}$ *lies in* $\mathbf{S}$*. That is,* $\mathbf{S}$ *is affine iff* $\mathbf{S} = $ *aff* $\mathbf{S}$*.*

Geometric interpretation of an affine set in $\mathbb{R}^n$:

A set $\mathbf{S}$ is affine if whenever $p, q \in S$ the entire line through points p, q is in the set $\mathbf{S}$.

At this point, we elaborate the close connection of affine sets and subspaces of $\mathbb{R}^n$. First we provide terminology for affine sets that emphasizes their relation with subspace of $\mathbb{R}^n$.

Definition 4.4. Flat in $\mathbb{R}^n$

A translate of a set $\mathbf{S}$ *in* $\mathbb{R}^n$ *by a vector* $\mathbf{p}$ *is the set* $\mathbf{S} + \mathbf{p} = \{\mathbf{s} + \mathbf{p}/\mathbf{s} \in \mathbf{S}\}$*. A* **flat** *in* $\mathbb{R}^n$ *is a translate of a subspace of* $\mathbb{R}^n$*.*

We note the following important points about a flat in $\mathbb{R}^n$*.*

- *Two flats are* **parallel** *if one is a translate of the other.*

- *The **dimension of a set** S, written as $\dim S$, is the dimension of the smallest flat conataning S.*

- *A **line** in $\mathbb{R}^n$ is a flat of dimension 1.*

- *A **hyperplane** in $\mathbb{R}^n$ is a flat of **dimension** $n-1$.*

Example 4.3. *State flats in $\mathbb{R}^3$.*

Solution: We know that proper subspaces of $\mathbb{R}^3$ are **the origin** O, **the set of all lines through** O, **the set of all planes through** O. The dimensions of these subspaces are **zero, one and two** respectively. From definition of a flat, it is clear that a translate of the subspaces are flats in $\mathbb{R}^3$. Thus the proper flats in $\mathbb{R}^3$ are **points (zero-dimensional), lines (one-dimensional) and planes (two-dimensional) which may or may not pass through the origin.**

The next theorem characterizes a set as affine set iff it is a flat.

Theorem 4.3. *A non-empty set $\mathbf{S}$ is a affine iff it is a flat.*

Example 4.4. *Illustrates theorem 4.3 geometrically. See figure below. The set of all linear comabination of vectors $\mathbf{b_1}, \mathbf{b_2}, \mathbf{b_3}$ is all of $\mathbb{R}^3$, but their **affine combination** is only a plane through b_1, b_2 and b_3 (that is plane $\mathbf{P_2}$) while P_1 is not in the plane.*

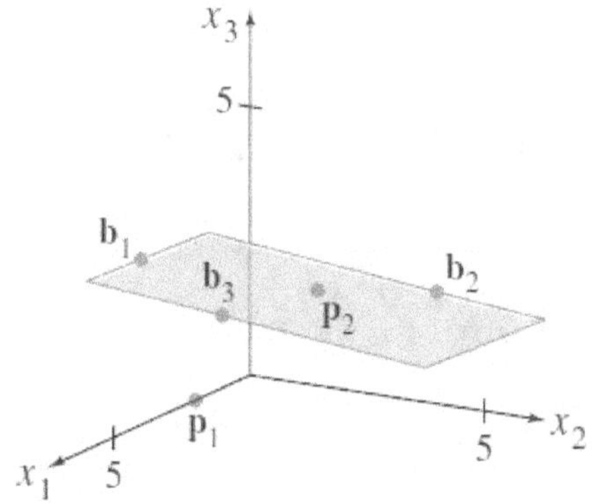

Figure 4.2

*So the theorem provides a geometric way to view the **affine hull** of a set ; It is nothing but a flat that consists of all the affine comabination of points in the set.*

*The next example explaines "**How one can express the solution set of the linear system** $Ax = b$ **as a set of affine comabination of certain vectors.**"*

Example 4.5. *Given a solution of a linear system* $\mathbf{Ax} = \mathbf{b}$*, say,*
$$\begin{bmatrix} x_1 \\ x_2 \\ x_3 \end{bmatrix} = \begin{bmatrix} 2x_3 \\ -3x_3 \\ x_3 \end{bmatrix} + \begin{bmatrix} 4 \\ 0 \\ -3 \end{bmatrix}$$

That is $\mathbf{x} = \mathbf{x_3 u} + \mathbf{P}$ *Where* $u = \begin{bmatrix} 2 \\ -3 \\ 1 \end{bmatrix}$ *and* $P = \begin{bmatrix} 4 \\ 0 \\ -3 \end{bmatrix}$. *Find the points* $\mathbf{v_1}$ *and* $\mathbf{v_2}$ *such that.*

Solution: Note that if $x = x_3 u + P$ is a solution of $\mathbf{Ax} = \mathbf{b}$ then the set $\mathbf{x} = x_3\mathbf{u}$ is a parallel to the solution set of $\mathbf{Ax} = \mathbf{0}$. This set consists of all the points of the form $\mathbf{x} = x_3\mathbf{u}$.

We give particular values to 'x_3' as follows:

$$x_3 = 0, \quad \mathbf{x} = x_3\mathbf{u} + \mathbf{p} = \mathbf{p} = \mathbf{v_1} \text{ (say)}$$

$$x_3 = 1, \quad \mathbf{x} = x_3 u + p = \mathbf{u} + \mathbf{p} = v_2 \text{ (say)}$$

That is $\mathbf{v_1} = \begin{bmatrix} 4 \\ 0 \\ -3 \end{bmatrix}$ and $\mathbf{v_2} = \begin{bmatrix} 2 \\ -3 \\ 1 \end{bmatrix} + \begin{bmatrix} 4 \\ 0 \\ -3 \end{bmatrix} = \begin{bmatrix} 6 \\ -3 \\ -2 \end{bmatrix}$

Now we write affine combination of $\mathbf{v_1}$ and $\mathbf{v_2}$ as follows:

$$\mathbf{x} = (1 - x_3)\mathbf{v_1} + x_3\mathbf{v_2}$$

This shows that every solution $\mathbf{x}$ of the system $\mathbf{Ax} = \mathbf{b}$ can be written as an affine comabination of vectors $\mathbf{v_1}$ and $\mathbf{v_2}$. So, the solution set of $\mathbf{Ax} = \mathbf{b}$ is aff $\{\mathbf{v_1}, \mathbf{v_2}\}$.

Example 4.6. *Suppose that the solutions of an equation* $\mathbf{Ax} = \mathbf{b}$ *are all of the form* $\mathbf{x} = x_3\mathbf{u} + \mathbf{p}$, *where* $\mathbf{u} = \begin{bmatrix} 4 \\ -2 \end{bmatrix}$ *and* $\mathbf{p} = \begin{bmatrix} -3 \\ 0 \end{bmatrix}$. *Find points* $\mathbf{v_1}$ *and* $\mathbf{v_2}$ *such that the solution set of* $\mathbf{Ax} = \mathbf{b}$ *is* *aff* $\{\mathbf{v_1}, \mathbf{v_2}\}$.

Solution: Let $x_3 = 0$, So, $\mathbf{x} = x_3\mathbf{u} + \mathbf{p} = \mathbf{p}$, define $\mathbf{v_1} = \mathbf{p}$ and for $x_3 = 1$ define $\mathbf{v_2} = \mathbf{u} + \mathbf{p}$.

Then we write $\mathbf{x} = (1 - x_3)\mathbf{v_1} + x_3\,\mathbf{v_2}$

$\therefore$ The required vectors are

$$v_1 = \begin{bmatrix} -3 \\ 0 \end{bmatrix}, \ v_2 = \begin{bmatrix} 4 \\ -2 \end{bmatrix} + \begin{bmatrix} -3 \\ 0 \end{bmatrix} = \begin{bmatrix} 1 \\ -2 \end{bmatrix}$$

By giving different values to x_3 other solutions are possible.

Definition 4.5. A homogeneous form of a vector v in $\mathbb{R}^n$

Let $\mathbf{v} \in \mathbb{R}^n$, *then the standard* **homogeneous form** *of a vector* $\mathbf{v}$ *is the point* $\tilde{v} = \begin{bmatrix} v \\ 1 \end{bmatrix}$ *in* $\mathbb{R}^{n+1}$

The following theorem characterizes the affine property of a vector in $\mathbb{R}^n$.

Theorem 4.4. *A vector* $y \in \mathbb{R}^n$ *is an affine combination of* $v_1, v_2, \cdots v_k$ *in* $\mathbb{R}^n$ *iff the homogeneous form of* $\tilde{y}$ *is in span* $\{\tilde{v}_1, \tilde{v}_2 \cdots \tilde{v}_k\}$.

Proof: Given that $\mathbf{y} \in$ aff $\{\mathbf{v_1}, \mathbf{v_2}, \cdots \mathbf{v_k}\}$ that is

$$\mathbf{y} = c_1\,\mathbf{v_1} + c_2\,\mathbf{v_2} + \cdots + c_k\,\mathbf{v_k}$$

where $c_1 + c_2 + \cdots + c_k = 1$.

Then

$$\begin{bmatrix} y \\ 1 \end{bmatrix} = c_1 \begin{bmatrix} v_1 \\ 1 \end{bmatrix} + c_2 \begin{bmatrix} v_2 \\ 1 \end{bmatrix} + \cdots + c_k \begin{bmatrix} v_k \\ 1 \end{bmatrix}$$

This is true iff $\tilde{y} \in$ Span $\{\tilde{v}_1, \tilde{v}_2 \cdots \tilde{v}_k\}$.

Hence proved the theorem.

Example 4.7. *Determine whether a vector **P** is expressible as an affine comabination of vectors* **v₁, v₂, v₃.** *where*

$$\mathbf{v_1} = \begin{bmatrix} 3 \\ 1 \\ 1 \end{bmatrix}, \quad \mathbf{v_2} = \begin{bmatrix} 1 \\ 2 \\ 2 \end{bmatrix}, \quad \mathbf{v_3} = \begin{bmatrix} 1 \\ 7 \\ 1 \end{bmatrix}, \quad \mathbf{P} = \begin{bmatrix} 4 \\ 3 \\ 0 \end{bmatrix}$$

.

Solution: To apply theorem 4.4 we first write the homogeneous coordinates of given vectors as follows:

$$\widetilde{\mathbf{v_1}} = \begin{bmatrix} 3 \\ 1 \\ 1 \\ 1 \end{bmatrix}, \quad \widetilde{\mathbf{v_2}} = \begin{bmatrix} 1 \\ 2 \\ 2 \\ 1 \end{bmatrix}, \quad \widetilde{\mathbf{v_3}} = \begin{bmatrix} 1 \\ 7 \\ 1 \\ 1 \end{bmatrix}, \quad \widetilde{\mathbf{P}} = \begin{bmatrix} 4 \\ 3 \\ 0 \\ 1 \end{bmatrix}$$

. Now, row reduce the augmented matrix for the equation

$$c_1\widetilde{v_1} + c_2\widetilde{v_2} + c_3\widetilde{v_3} = \widetilde{P} \text{ as } [\widetilde{v_1}\ \widetilde{v_2}\ \widetilde{v_3}\ \widetilde{P}]$$

$$\sim \begin{bmatrix} 3 & 1 & 1 & 4 \\ 1 & 2 & 7 & 3 \\ 1 & 2 & 1 & 0 \\ 1 & 1 & 1 & 1 \end{bmatrix} \sim \begin{bmatrix} 1 & 1 & 1 & 1 \\ 3 & 1 & 1 & 4 \\ 1 & 2 & 7 & 3 \\ 1 & 2 & 1 & 0 \end{bmatrix} \sim \begin{bmatrix} 1 & 0 & 0 & 1\cdot 5 \\ 0 & 1 & 0 & -1 \\ 0 & 0 & 1 & 0.5 \\ 0 & 0 & 0 & 0 \end{bmatrix}$$

By theorem (4.4) we write

$$1\cdot 5\mathbf{v_1} - \mathbf{v_2} + 0\cdot 5\mathbf{v_3} = \mathbf{P}$$

This is the required affine combination of **P** in terms of **v₁, v₂, v₃**.

The following figure expresses it geometrically.

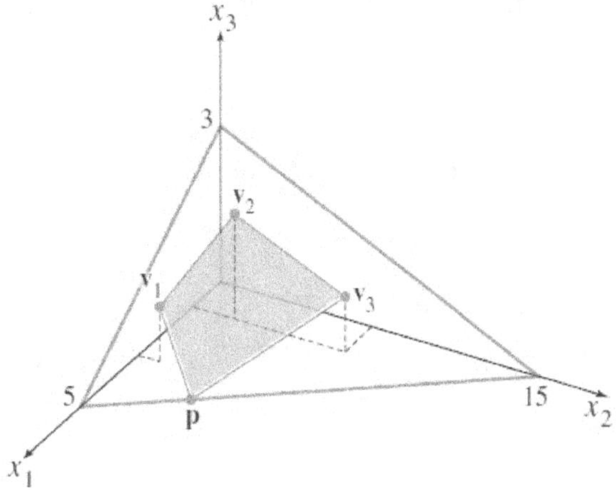

Figure 4.3

Example 4.8. *Find the affine combination for **P** using vectors **v₁, v₂, v₃**. Draw the graph. Given*

$$\mathbf{v_1} = \begin{bmatrix} 1 \\ 0 \end{bmatrix}, \quad \mathbf{v_2} = \begin{bmatrix} -1 \\ 2 \end{bmatrix}, \quad \mathbf{v_3} = \begin{bmatrix} 3 \\ 1 \end{bmatrix}, \quad \mathbf{P} = \begin{bmatrix} 4 \\ 3 \end{bmatrix}$$

Solution: Write homogeneous coordinates of the given vectors:

$$\widetilde{v_1} = \begin{bmatrix} 1 \\ 0 \\ 1 \end{bmatrix} \quad \widetilde{v_2} = \begin{bmatrix} -1 \\ 2 \\ 1 \end{bmatrix} \quad \widetilde{v_3} = \begin{bmatrix} 3 \\ 1 \\ 1 \end{bmatrix} \quad \widetilde{P} = \begin{bmatrix} 4 \\ 3 \\ 1 \end{bmatrix}$$

Row reduce the augmented matrix for the equation

$$c_1\widetilde{v_1} + c_2\widetilde{v_2} + c_3\widetilde{v_3} = \widetilde{P}$$

$$[\widetilde{v_1}\ \widetilde{v_2}\ \widetilde{v_3}\ \widetilde{P}] \sim \begin{bmatrix} 1 & -1 & 3 & 4 \\ 0 & 2 & 1 & 3 \\ 1 & 1 & 1 & 1 \end{bmatrix} \sim \begin{bmatrix} 1 & 1 & 1 & 1 \\ 1 & -1 & 3 & 4 \\ 0 & 2 & 1 & 3 \end{bmatrix} \sim \begin{bmatrix} 1 & 0 & 0 & \frac{-3}{2} \\ 0 & 1 & 0 & \frac{1}{2} \\ 0 & 0 & 1 & 2 \end{bmatrix}$$

Hence $-\frac{3}{2}v_1 + \frac{1}{2}v_2 + 2v_3 = P$ By theorem 4.4.

4.3   Affine independence

Let $\mathbf{S}$ be a set of vectors of $\mathbb{R}^3$, $\mathbf{v_1}, \mathbf{v_2}, \mathbf{v_3}$ say. Then the set is $\mathbf{S} = \{\mathbf{v_1}, \mathbf{v_2}, \mathbf{v_3}\}$ is an affine combination means one of the vectors of $\mathbf{S}$ is an affine combination of other vectors.

That is, suppose

$$v_3 = (1 - t)v_1 + tv_2, \ t \in \mathbb{R}$$

Then

$$(1 - t)v_1 + tv_2 - v_3 = 0$$

From defination of linear dependence we know that this relation is linearly dependent as **not** all coefficients are zero. However, we note that the sum of these coefficients is 0 as given below

$$(1 - t) + t - 1 = 0$$

This is an additional property needed to define affine dependence. We now define affine dependence of vectors as follows:

Definition 4.6. Affinely dependent set

An indexed set of points vectors $\{v_1, v_2, \cdots v_k\}$ in $\mathbb{R}^n$, is **affinely dependent** *if there exist real numbers* $\mathbf{c_1}, \mathbf{c_2}, \cdots \mathbf{c_k}$ **not** *all zero, Such that*

$$c_1 + c_2 + \cdots + c_k = D \ [and\ ]c_1v_1 + v_2v_2 \cdots + c_kv_k = 0$$

otherwise the set is affinely independent.

Example 4.9. *Following sets are examples of affinely dependent / independent sets.*

Solution:

(b) The affine hull of two distinct points **p** and **q** is a line l. Let r be a third point on this line. Then the set $\{p, q, r\}$ is an **affinely** dependent set.

(c) If another point s is **not** on the line l then the set $\{p, q, s\}$ is **an affinely independent set.**

(d) An indexed set $\{\mathbf{v_1}, \mathbf{v_2}\}$ in $\mathbb{R}^n$ is affinely dependent iff $\mathbf{v_1} = \mathbf{v_2}$.

From definition of affine combination of vectors it is clear that it is a special type of linear combination of vectors. Moreover, affine dependence of vectors is a restricted type of linear dependence. Thus each affinely dependent set is automatically linearly dependent.

The following theorem shows how the concept of affine dependence is analogous to linear dependence. The theorem also provide the useful method to determine whether a set is affinely dependent.

Theorem 4.5. *Given an indexed set* $\mathbf{s} = \{\mathbf{v_1}, \mathbf{v_2}, \cdots \mathbf{v_p}\}$ *in* $\mathbb{R}^n$, *with* $p \geq 2$, *the following statement are logically equivalent.*

(a) **S** *is affinely dependent.*

(b) *One of the points in* **S** *is an affine combination of the other points in* **S**.

(c) *The set* $\{\mathbf{v_2} - \mathbf{v_1}, \cdots \mathbf{v_p} - \mathbf{v_1}\}$ *in* $\mathbb{R}^n$ *is linearly dependent.*

d) *The set* $\{\widetilde{\mathbf{v_1}}, \widetilde{\mathbf{v_2}}, \cdots \widetilde{\mathbf{v_p}}\}$ *of homogeneous forms in* $\mathbb{R}^{n+1}$ *is linearly dependent.*

Example 4.10. *Determine whether the set* **S** *given below is affinely independent.*

$$s = \{v_1, v_2, v_3\} \quad \text{where} \quad v_1 = \begin{bmatrix} 1 \\ 3 \\ 7 \end{bmatrix}, \ v_2 = \begin{bmatrix} 2 \\ 7 \\ 6.5 \end{bmatrix}, \ v_3 = \begin{bmatrix} 0 \\ 4 \\ 7 \end{bmatrix}$$

Solution: Compute $v_2 - v_1$ and $v_3 - v_1$.

$$\mathbf{v_2} - \mathbf{v_1} = \begin{bmatrix} 2 \\ 7 \\ 6.5 \end{bmatrix} - \begin{bmatrix} 1 \\ 3 \\ 7 \end{bmatrix} = \begin{bmatrix} 1 \\ 4 \\ -0.5 \end{bmatrix}$$

$$\mathbf{v_3} - \mathbf{v_1} = \begin{bmatrix} 0 \\ 4 \\ 7 \end{bmatrix} - \begin{bmatrix} 1 \\ 3 \\ 7 \end{bmatrix} = \begin{bmatrix} -1 \\ 1 \\ 0 \end{bmatrix}$$

The vectors $\mathbf{v_2} - \mathbf{v_1}$ and $\mathbf{v_3} - \mathbf{v_1}$ are **not** multiples of each other. Hence, the vectors $v_2 - v_1, \ v_3 - v_1$ forms linearly independent set.

Hence, by theorem (4.5) the given set **S** is affinely independent.

The following figure shows the span of $\{v_2 - v_1, \ v_3 - v_1\}$ and set **S**. Note that both the planes are parallel to each other. One passing through origin.

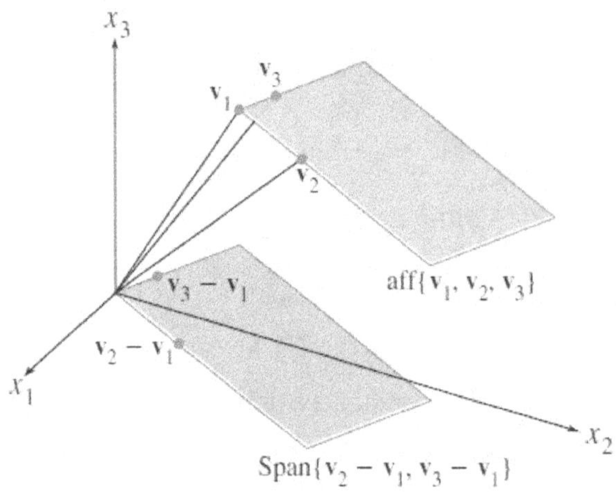

Figure 4.4

The following example explains the method to determine whether the given set of vectors is affinely dependent or not.

Example 4.11. *Given set* $\mathbf{S} = \{\mathbf{v_1}, \mathbf{v_2}, \mathbf{v_3}, \mathbf{v_4}\}$ *where*

$$v_1 = \begin{bmatrix} 1 \\ 3 \\ 7 \end{bmatrix}, \quad v_2 = \begin{bmatrix} 2 \\ 7 \\ 6.5 \end{bmatrix}, \quad v_3 = \begin{bmatrix} 0 \\ 4 \\ 7 \end{bmatrix}, \quad v_4 = \begin{bmatrix} 0 \\ 14 \\ 6 \end{bmatrix}$$

Determine whether the set $\mathbf{S}$ *is affinely dependent.*

Solution: Step 1:

Compute the following vectors:

$v_4 - v_1, v_3 - v_1, v_2 - v_1$ as follows.

$$v_2 - v_1 = \begin{bmatrix} 1 \\ 4 \\ -\cdot 5 \end{bmatrix}, \quad v_3 - v_1 = \begin{bmatrix} -1 \\ 1 \\ 0 \end{bmatrix}, \quad v_4 - v_1 = \begin{bmatrix} -1 \\ 11 \\ -1 \end{bmatrix},$$

$$[v_2 - v_1, v_3 - v_1, v_4 - v_1] = \begin{bmatrix} 1 & -1 & -1 \\ 4 & 1 & 11 \\ -0 \cdot 5 & 0 & -1 \end{bmatrix}$$

Step 2: Row reduce the above matrix.

$$\begin{bmatrix} 1 & -1 & -1 \\ \cdot 4 & 1 & 11 \\ -0 \cdot 5 & 0 & -1 \end{bmatrix} \sim \begin{bmatrix} 1 & -1 & -1 \\ 4 & 5 & 15 \\ 0 & 0 & 0 \end{bmatrix}$$

Note that there are only two pivot elements. So the set $[v_2 - v_1, v_3 - v_1, v_4 - v_1]$ is linearly dependent.

Hence by statement *c*) of theorem 4.5 the given set S is affinely dependent.

Actually, the matrix $\begin{bmatrix} 1 & -1 & -1 \\ 0 & 5 & 15 \\ 0 & 0 & 0 \end{bmatrix}$ from above shows that vector $\mathbf{v_4} - \mathbf{v_1}$ can be written as a linear combination of vectors $v_2 - v_1, v_3 - v_1$.

$$\begin{bmatrix} 1 & -1 & -1 \\ 0 & 5 & 15 \\ 0 & 0 & 0 \end{bmatrix} \sim \begin{bmatrix} 1 & 0 & 2 \\ 0 & 1 & 3 \\ 0 & 0 & 0 \end{bmatrix}$$

The above row-reduced matrix shows that $2(v_2 - v_1) + 3(v_3 - v_1) = v_4 - v_1$ Simplifying the above equation. we get $\mathbf{v_4} = -4\mathbf{v_1} + 2\mathbf{v_2} + 3\mathbf{v_3}$. This shows that $\mathbf{v_4}$ is expressed as a linear combination of vectors v_1, v_2 and v_3. In fact it is an affine combination; as $-4 + 2 + 3 = 1$. Hence the set $\{v_1, v_2, v_3, v_4\}$ is affine dependent and this implies that set $\mathbf{S}$ is linearly dependent.

Geometric visualization of the above example is given in Figure below

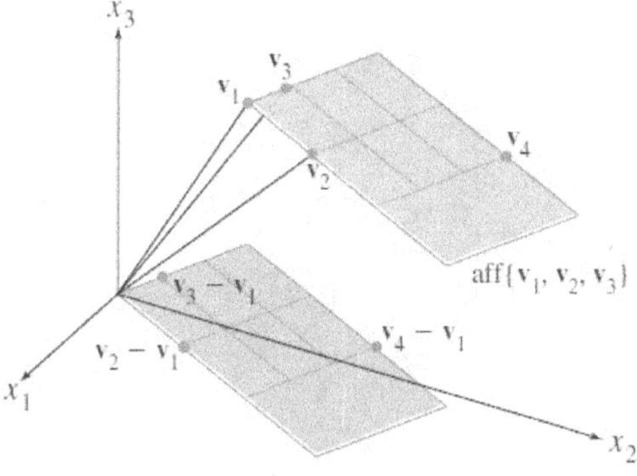

Figure 4.5

In the figure we see two planes. One plane represents Span $\{v_2 - v_1, v_3 - v_1\}$ and the other is aff $\{v_1, v_2, v_3\}$. The coordinate system based on the

$$v_4 = -4v_1 + 2v_2 + 3v_3$$

in which the coefficients are $-4, 2, 3$ and v_4 is in the plane $\mathbf{aff}\ \{\mathbf{v_1}, \mathbf{v_2}, \mathbf{v_3}\}$ refers to **Barycentric coordinates or affine coordinates** of a vector $\mathbf{v_4}$. Barycentric coordinates of a vector are useful in **Computer Graphics** to color, to shade small surface as per the demand and requirement.

4.4   Convex Combinations

In section, $4 \cdot 1$ we studied a special type of linear comabination known as "Affine Combination" of the form $c_1 v_1 + c_2 v_2 + \cdots + c_k v_k$ such that $c_1 + c_2 + \cdots + c_k = 1$. In this section we restric the values $c_1, c_2, \cdots c_k$ further to be **non-negative**.

Such a comabination of vectors is known as "**Convex combination of vectors.**"

Definition 4.7. *A convex combination of points*

Let $v_1, v_2 \cdots v_k \in \mathbb{R}^n$. Consider linear combinationof $v_1, v_2 \cdots v_k$ a follows:

$c_1 v_1 + c_2 v_2 + \cdots + c_k v_k$ such that $c_1 + c_2 + \cdots + c_k = 1$ and $c_i \geq 0\ \forall i$. The set of all convex

Example 4.12. *The following are some examples of a* **convex hull (conv S).**

Solution:

(a) $S = \{v_1\}$. The **conv** S of a singleton is the same as the **aff** S.

(b) $S = \{v_1, v_2\}$, $v_1 \neq v_2$. We know that aff S is **a single line** with the equation.

$$y = (1-t)v_1 + tv_2, \ t \in \mathbb{R}$$

In particular, for a convex combinational restrict values of t as $\mathbf{0 \leq t \leq 1}$. It is a line segment between v_1 and v_2. **Here convex hull S is properly contained in affine hull of S.**

Example 4.13. *Given an orthogonal set* $\mathbf{S} = \{\mathbf{v_1}, \mathbf{v_2}, \mathbf{v_3}\}$ *where*

$$v_1 = \begin{bmatrix} 3 \\ 0 \\ 6 \\ -3 \end{bmatrix}, \ v_2 = \begin{bmatrix} -6 \\ 3 \\ 3 \\ 0 \end{bmatrix}, \ v_3 = \begin{bmatrix} 3 \\ 6 \\ 0 \\ 3 \end{bmatrix}$$

Determine whether the vector $\mathbf{P_1}$ *is in conv* $\mathbf{S}$.

Solution: Given $P_1 = \begin{bmatrix} 0 \\ 3 \\ 3 \\ 0 \end{bmatrix}$. Write the matrix $[v_1 \ v_2 \ v_3 \ P_1]$, it is an augmented matrix. Row reduce the matrix as follows

$$\begin{bmatrix} 3 & -6 & 3 & 0 \\ 0 & 3 & 6 & 3 \\ 6 & 3 & 0 & 3 \\ -3 & 0 & 3 & 0 \end{bmatrix} \sim \begin{bmatrix} 1 & -2 & 1 & 0 \\ 0 & 1 & 2 & 1 \\ 2 & 1 & 0 & 1 \\ 1 & 0 & -1 & 0 \end{bmatrix} \sim \begin{bmatrix} 1 & -2 & 1 & 0 \\ 0 & 1 & 2 & 1 \\ 0 & 5 & -2 & 1 \\ 0 & 2 & -2 & 0 \end{bmatrix}$$

$$\sim \cdots \sim \begin{bmatrix} 1 & -2 & 0 & \frac{-1}{3} \\ 0 & 1 & 0 & \frac{1}{3} \\ 0 & 0 & 1 & \frac{1}{3} \\ 0 & 0 & 0 & 0 \end{bmatrix} \sim \begin{bmatrix} 1 & 0 & 0 & \frac{1}{3} \\ 0 & 1 & 0 & \frac{1}{3} \\ 0 & 0 & 1 & \frac{1}{3} \\ 0 & 0 & 0 & 0 \end{bmatrix}$$

The fourth column shows that

$$P_1 = \frac{1}{3}v_1 + \frac{1}{3}v_2 + \frac{1}{3}v_3$$

Note that $\frac{1}{3} + \frac{1}{3} + \frac{1}{3} = 1$ and the weight (coefficients) are non-negative. **Hence $P_1 \in$ aff** S and $P_1 \in$ **conv** S.

Example 4.14. *For the same set of vectors S given example 4.13, Determine whether the vector* $\mathbf{P_2}$ *given below is in conv* S.

Solution: We know that the given set S is an orthogonal set. We find orthogonal projection of $\mathbf{P_2}$ onto a subspace spanned by set S as follows:

Given $P_2 = \begin{bmatrix} -10 \\ 5 \\ 11 \\ -4 \end{bmatrix}$.

$$\operatorname{proj}_w P_2 = \frac{P_2 \cdot v_1}{v_1 \cdot v_1} v_1 + \frac{P_2 \cdot v_2}{v_2 \cdot v_2} v_2 + \frac{P_2 \cdot v_3}{v_3 \cdot v_3} v_3$$

$$= \frac{48}{54} v_1 + \frac{108}{54} v_2 + \frac{-12}{54} v_3$$

$$= \frac{8}{9} v_1 + 2 v_2 + \frac{-2}{9} v_3$$

$$= \begin{bmatrix} -10 \\ \frac{14}{3} \\ \frac{34}{3} \\ \frac{-10}{3} \end{bmatrix} \neq P_2$$

This shows that $\mathbf{P_2} \notin$ Span $\mathbf{S}$. In particular, $\mathbf{P_2} \notin$ aff $\mathbf{S}$ hence $\mathbf{P_2} \notin$ conv $\mathbf{S}$.

The geometric interpretation of an affine set S is thet it contains all lines determined by pair of points in S.

When we restrict the affine comabination of vectors to convex comabination of vectors the appropriate condition involves line segments rather than lines. This is shown in the following figure.

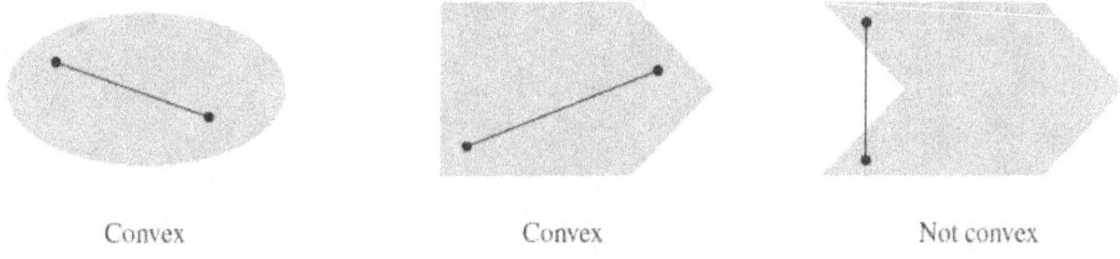

Figure 4.6

Definition 4.8. Convex Set

A set $\mathbf{S}$ *is said to be convex if for each* $\mathbf{p}, \mathbf{q} \in \mathbf{S}$ *the line segment* $\overline{\mathbf{pq}}$ *is contained in* $\mathbf{S}$.

In other words, in a convex set every two points in a set can "**see**" **each other without the line of sight leaving the set.**

The following theorem is a characterization of a convex set.

Theorem 4.6. *A set* $\mathbf{S}$ *is convex iff every convex comabination of points of* $\mathbf{S}$ *lies in* $\mathbf{S}$*. That is,* $\mathbf{S}$ *is convex iff* $\mathbf{S} = conv\ \mathbf{S}$.

Properties of convex sets and affine sets:

1. If any collection of affine sets is given, then intersection of these affine sets is always an affine set.

2. If any collection of convex sets is given, then intersection of these convex sets is always a convex

3. For any set **S**, the convex hull of **S** is the intersection of all convex sets that contain **S**. That is the convex hull of **S** is the **smallest** convex set that contains **S**.

4. In particular, in $\mathbb{R}^2$, the convex hull of set **S Fill in** all the holes in the inside of S and **fill out** all the dents in the boundary of S. See figure below.

5. Let **S** be a given set in $\mathbb{R}^3$. **S** $= \{\mathbf{e_1}, \mathbf{e_2}, \mathbf{e_3}\}$ as shown in the following figure. Then conv **S** (convex hull of S) is a triangular surface in $\mathbb{R}^3$.

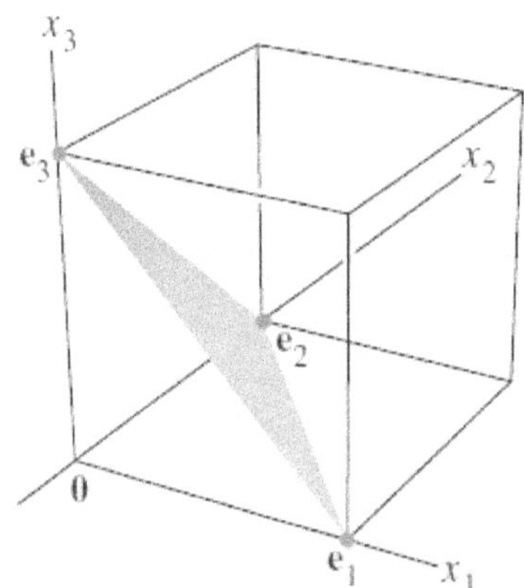

The properties listed above are very useful in solving problems on affine sets and convex sets. The following theorem tells us 'how to express a point in convex hull of S whenever S is any non-empty subset of $\mathbb{R}^{n}$'

Theorem 4.7. (Caratheodory)

If **S** *is a non-empty subset of* $\mathbb{R}^n$, *then every point in* **conv** S *(convex hull of S) can be expressed a a convex combination of $n+1$ or fewer points of* **S**.

Example 4.15. *Given set* $S = \left\{ \begin{bmatrix} x \\ y \end{bmatrix} \middle/ x \geq 0 \text{ and } y = x^2 \right\}$. *Show that the convex hull of* **S** *is the union of the origin and* $\left\{ \begin{bmatrix} x \\ y \end{bmatrix} \middle/ x > 0 \text{ and } y \geq x^2 \right\}$.

Solution: We know that, every point in convex hull of S must lie on a line segment joining two points of S. Refer to the following figure.

The dashed $y-$ axis in the figure shows that the positive $y-$ axis is **not** in convex hull S; because

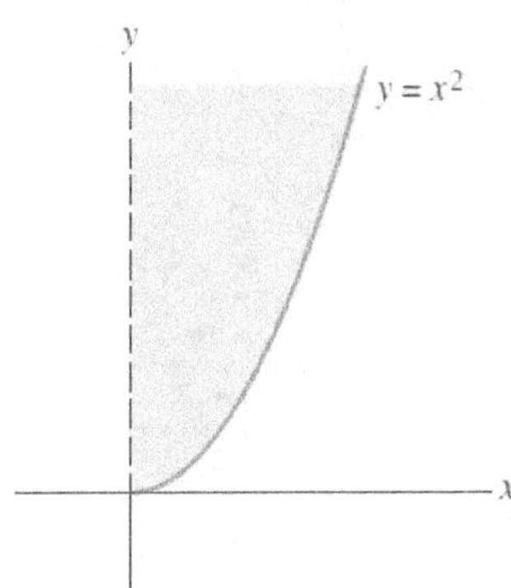

and the set $\left\{\begin{bmatrix} x \\ y \end{bmatrix} / x > 0 \text{ and } y \geq x^2 \right\}$. Now, we show analytically that any point $P = \begin{bmatrix} a \\ b \end{bmatrix}$ in the set S is on a line segment from the origin to a point on the curve in S.

Let $P = \begin{bmatrix} a \\ b \end{bmatrix}$ be in the shaded portion of above figure, then from the given conditions of the region we see that $a > 0$ and $b \geq a^2$.

The equation of a line through the origin $\mathbf{O}$ and point $\mathbf{P}$ is $y = (\frac{a}{b})t$ for $t \in \mathbb{R}$. This line intersects S where t satisfies $(\frac{b}{a})t = t^2$ that is when $t = \frac{b}{a}$.

Thus P in on the line segment joining the origin and the point with coordinates $\begin{bmatrix} \frac{b}{a} \\ \frac{b^2}{a^2} \end{bmatrix}$.

EXERCISE

1. Write $\mathbf{y}$ as an affine combination of the other points listed, if possible.

(a) $v_1 = \begin{bmatrix} 1 \\ 2 \end{bmatrix}$, $v_2 = \begin{bmatrix} -2 \\ 2 \end{bmatrix}$, $v_3 = \begin{bmatrix} 0 \\ 4 \end{bmatrix}$, $v_4 = \begin{bmatrix} 3 \\ 7 \end{bmatrix}$, $y = \begin{bmatrix} 5 \\ 3 \end{bmatrix}$

(b) $v_1 = \begin{bmatrix} 1 \\ 1 \end{bmatrix}$, $v_2 = \begin{bmatrix} -1 \\ 2 \end{bmatrix}$, $v_3 = \begin{bmatrix} 3 \\ 2 \end{bmatrix}$, $y = \begin{bmatrix} 5 \\ 7 \end{bmatrix}$

(c) $v_1 = \begin{bmatrix} -3 \\ 1 \\ 1 \end{bmatrix}$, $v_2 = \begin{bmatrix} 0 \\ 4 \\ -2 \end{bmatrix}$, $v_3 = \begin{bmatrix} 4 \\ -2 \\ 6 \end{bmatrix}$, $y = \begin{bmatrix} 17 \\ 1 \\ 5 \end{bmatrix}$

(d) $v_1 = \begin{bmatrix} 1 \\ 2 \\ 0 \end{bmatrix}$, $v_2 = \begin{bmatrix} 2 \\ -6 \\ 7 \end{bmatrix}$, $v_3 = \begin{bmatrix} 4 \\ 3 \\ 1 \end{bmatrix}$, $y = \begin{bmatrix} -3 \\ 4 \\ -4 \end{bmatrix}$

.2 Let $S = \{b_1, b_2, b_3\}$ be an orthogonal basis for $\mathbb{R}^3$. Write each of the the given points as an affine combination of the points in the set S, if possible.

(a) $b_1 = \begin{bmatrix} 2 \\ 1 \\ 1 \end{bmatrix}$, $b_2 = \begin{bmatrix} 1 \\ 0 \\ -2 \end{bmatrix}$, $b_3 = \begin{bmatrix} 2 \\ -5 \\ 1 \end{bmatrix}$

$b_1 = \begin{bmatrix} 3 \\ 8 \\ 4 \end{bmatrix}$, $b_2 = \begin{bmatrix} 6 \\ -3 \\ 3 \end{bmatrix}$, $b_3 = \begin{bmatrix} 0 \\ -1 \\ -5 \end{bmatrix}$

(b) $b_1 = \begin{bmatrix} 2 \\ 1 \\ 1 \end{bmatrix}$, $b_2 = \begin{bmatrix} 1 \\ 0 \\ -2 \end{bmatrix}$, $b_3 = \begin{bmatrix} 2 \\ -5 \\ 1 \end{bmatrix}$

$b_1 = \begin{bmatrix} 0 \\ -19 \\ -5 \end{bmatrix}$, $b_2 = \begin{bmatrix} 1 \cdot 5 \\ -1 \cdot 3 \\ -0 \cdot 5 \end{bmatrix}$, $b_3 = \begin{bmatrix} 5 \\ -4 \\ 0 \end{bmatrix}$

3. Suppose that the solution of an equal $\mathbf{Ax} = \mathbf{b}$ are all of the form $\mathbf{x} = x, \mathbf{u_1} + \mathbf{p}$, where

$\mathbf{u} = \begin{bmatrix} 5 \\ 1 \\ -2 \end{bmatrix}$, $\mathbf{P} = \begin{bmatrix} 1 \\ -3 \\ 4 \end{bmatrix}$. Find points $\mathbf{v_1}$ and $\mathbf{v_2}$ such that the solution set of $\mathbf{Ax} = \mathbf{b}$ is aff $\{\mathbf{v_1}, \mathbf{v_2}\}$.

4. State true or false. Justify each answer.

(a) The set of all affine combinations of points in a set $\mathbf{S}$ is called the affine hull of $\mathbf{S}$.

(b) A **flat** is a subspace.

(c) A plane in $\mathbb{R}^3$ is a hyperplane.

(d) If $\mathbf{S} = \{\mathbf{x}\}$, then aff bfS is the empty set.

(e) A flat of dim 1 is called a line.

5. Let $\mathbf{S} = \{\mathbf{v_1}, \mathbf{v_2}, \mathbf{v_3}\}$, and given vectors P_1, P_2, P_3. Determine whether vectors P_1, P_2, P_3 are expressible as affine combination of vectors of $\mathbf{S}$. Where,

(a) $\mathbf{v_1} = \begin{bmatrix} 1 \\ 0 \\ 3 \\ 0 \end{bmatrix}$, $\mathbf{v_2} = \begin{bmatrix} 2 \\ -1 \\ 0 \\ 4 \end{bmatrix}$, $\mathbf{v_3} = \begin{bmatrix} -1 \\ 2 \\ 1 \\ 1 \end{bmatrix}$

$\mathbf{P_1} = \begin{bmatrix} 5 \\ -3 \\ 5 \\ 3 \end{bmatrix}$, $\mathbf{P_2} = \begin{bmatrix} -9 \\ 10 \\ 9 \\ -13 \end{bmatrix}$, $\mathbf{P_3} = \begin{bmatrix} 4 \\ 2 \\ 8 \\ 5 \end{bmatrix}$

(b) $\mathbf{v_1} = \begin{bmatrix} 1 \\ 0 \\ 3 \\ -2 \end{bmatrix}$, $\mathbf{v_2} = \begin{bmatrix} 2 \\ 1 \\ 6 \\ -5 \end{bmatrix}$, $\mathbf{v_3} = \begin{bmatrix} 3 \\ 0 \\ 12 \\ -6 \end{bmatrix}$

$\mathbf{P_1} = \begin{bmatrix} 4 \\ -1 \\ 15 \\ -7 \end{bmatrix}$, $\mathbf{P_2} = \begin{bmatrix} -5 \\ 3 \\ -8 \\ 6 \end{bmatrix}$, $\mathbf{P_3} = \begin{bmatrix} 1 \\ 6 \\ -6 \\ -8 \end{bmatrix}$

6. Determine if the set of points is affinely dependent, if so, construct an affine dependence relation for the points.

(a) $\begin{bmatrix} 3 \\ -3 \end{bmatrix}, \begin{bmatrix} 0 \\ 6 \end{bmatrix}, \begin{bmatrix} 2 \\ 0 \end{bmatrix}$ (b) $\begin{bmatrix} 2 \\ 1 \end{bmatrix}, \begin{bmatrix} 5 \\ 4 \end{bmatrix}, \begin{bmatrix} -3 \\ -2 \end{bmatrix}$

(c) $\begin{bmatrix} 1 \\ 2 \end{bmatrix}, \begin{bmatrix} -2 \\ -3 \end{bmatrix}, \begin{bmatrix} 2 \\ -1 \end{bmatrix}, \begin{bmatrix} 0 \\ 15 \end{bmatrix}$ (d) $\begin{bmatrix} -2 \\ 5 \end{bmatrix}, \begin{bmatrix} 0 \\ -3 \end{bmatrix}, \begin{bmatrix} 1 \\ -2 \end{bmatrix}, \begin{bmatrix} -2 \\ 7 \end{bmatrix}$

(e) $\begin{bmatrix} 1 \\ 0 \\ -2 \end{bmatrix}$, $\begin{bmatrix} 0 \\ 1 \\ 1 \end{bmatrix}$, $\begin{bmatrix} -1 \\ 5 \\ 1 \end{bmatrix}$, $\begin{bmatrix} 0 \\ 5 \\ -3 \end{bmatrix}$ $\qquad$ (f) $\begin{bmatrix} 1 \\ 3 \\ 1 \end{bmatrix}$, $\begin{bmatrix} 0 \\ -1 \\ -2 \end{bmatrix}$, $\begin{bmatrix} 2 \\ 5 \\ 2 \end{bmatrix}$, $\begin{bmatrix} 3 \\ 5 \\ 0 \end{bmatrix}$

7. Let $v_1 = \begin{bmatrix} 0 \\ 1 \end{bmatrix}$, $v_2 \begin{bmatrix} 1 \\ 5 \end{bmatrix}$, $v_3 = \begin{bmatrix} 4 \\ 3 \end{bmatrix}$

   Show that the set S is affinely independent.

8. Show that the points $v_1 = \begin{bmatrix} 4 \\ 1 \end{bmatrix}$, $v_2 \begin{bmatrix} 1 \\ 0 \end{bmatrix}$, $v_3 = \begin{bmatrix} 5 \\ 4 \end{bmatrix}$, $v_4 = \begin{bmatrix} 5 \\ 4 \end{bmatrix}$ form an affinely dependent set.

9. Let $v_1 = \begin{bmatrix} 6 \\ 2 \\ 2 \end{bmatrix}$, $v_2 \begin{bmatrix} 7 \\ 1 \\ 5 \end{bmatrix}$, $v_3 = \begin{bmatrix} -2 \\ 4 \\ -1 \end{bmatrix}$, $P_1 = \begin{bmatrix} 1 \\ 3 \\ 1 \end{bmatrix}$, $P_2 = \begin{bmatrix} 3 \\ 2 \\ 1 \end{bmatrix}$ and let $\mathbf{S} = \{\mathbf{v_1, v_2, v_3}\}$.

   Determine whether P_1 and P_2 are in conv $\mathbf{S}$.

10. Let $S = \{v_1, v_2, v_3, v_4\}$ where

$$v_1 = \begin{bmatrix} -1 \\ -3 \\ 4 \end{bmatrix}, \quad v_2 \begin{bmatrix} 0 \\ -3 \\ 1 \end{bmatrix}, \quad v_3 = \begin{bmatrix} 1 \\ -1 \\ 4 \end{bmatrix}, \quad v_4 = \begin{bmatrix} 1 \\ 1 \\ -2 \end{bmatrix}$$

   Determine whether points P_1 and P_2 are in conv S; where $P_1 = \begin{bmatrix} 1 \\ -1 \\ 2 \end{bmatrix}$, $P_2 = \begin{bmatrix} 0 \\ -2 \\ 2 \end{bmatrix}$.

11. Let $S = \{v_1, v_2, v_3\}$ where

$$v_1 = \begin{bmatrix} 2 \\ 0 \\ -1 \\ 2 \end{bmatrix}, \quad v_2 \begin{bmatrix} 0 \\ -2 \\ 2 \\ 1 \end{bmatrix}, \quad v_3 = \begin{bmatrix} -2 \\ 1 \\ 0 \\ 2 \end{bmatrix}$$

   Determine whether points P_1 and P_2 are in conv. S; where $P_1 = \begin{bmatrix} -1 \\ 2 \\ \frac{-3}{2} \\ \frac{5}{2} \end{bmatrix}$, $P_2 = \begin{bmatrix} \frac{-1}{2} \\ 0 \\ \frac{1}{4} \\ \frac{7}{4} \end{bmatrix}$.

12. Let S. be the set of points on the curve $y = \frac{1}{x}$ for $x > 0$. Explain geometrically why **conv** S consists of all points on and above the curve S.

13. In $\mathbb{R}^2$, let $\mathbf{S} = \left\{ \begin{bmatrix} 0 \\ y \end{bmatrix} / 0 \le y < 1 \right\} \cup \left\{ \begin{bmatrix} 2 \\ 0 \end{bmatrix} \right\}$.

   Sketch the convex hull of $\mathbf{S}$.

14. Show that the convex hull of set S of points $\begin{bmatrix} x \\ y \end{bmatrix}$ in $\mathbb{R}^2$ that satisfy the given conditions. Justify your answer.

   Show that an arbitrary point $\mathbf{P}$ in S belongs to conv S.

   (a) $y = \frac{1}{x}$ and $x \ge \frac{1}{2}$ $\qquad$ (b) $y = \sin x$ $\qquad$ (c) $y = x^{\frac{1}{2}}$ and $x \ge 0$.

15. Let $\mathbf{S} = \{\mathbf{b_1}, \mathbf{b_2}, \mathbf{b_3}\}$ where

$$\mathbf{b_1} = \begin{bmatrix} 2 \\ 1 \\ 1 \end{bmatrix}, \quad \mathbf{b_2} = \begin{bmatrix} 1 \\ 0 \\ -2 \end{bmatrix}, \quad \mathbf{b_3} = \begin{bmatrix} 2 \\ -5 \\ 1 \end{bmatrix}$$

which of the points P_1, P_2 and P_3 are in conv S? where,

$$\mathbf{P_1} = \begin{bmatrix} 3 \\ 8 \\ 4 \end{bmatrix}, \quad \mathbf{P_2} = \begin{bmatrix} 6 \\ -3 \\ 3 \end{bmatrix}, \quad \mathbf{P_3} = \begin{bmatrix} 0 \\ -1 \\ -5 \end{bmatrix}$$

16. Let S be a set of vectors given in problem 15. which of the points P_1, P_2 and P_3 are in conv S? where,

$$\mathbf{P_1} = \begin{bmatrix} 0 \\ -19 \\ -5 \end{bmatrix}, \quad \mathbf{P_2} = \begin{bmatrix} 1 \cdot 5 \\ -1 \cdot 3 \\ -0 \cdot 5 \end{bmatrix}, \quad \mathbf{P_3} = \begin{bmatrix} 5 \\ -4 \\ 0 \end{bmatrix}$$

SOLUTIONS

1. a) $y = 2.v_1 + 0.v_2 - 2.v_3 + v_4$ other possible answers $y = 2.v_1 + (\frac{-3}{2} + \frac{3}{2}r).v_2 + (\frac{1}{2} + \frac{-5}{2}r).v_3 + r.v_4$,

   r any real number

   b) $y = -5.v_1 + 2.v_2 + 4.v_3$

   c) $y = -3.v_1 + 2.v_2 + 2.v_3$

   d) $y = \frac{26}{10}.v_1 + \frac{-4}{10}.v_2 + \frac{-12}{10}.v_3$

2. a) $p_1 = 3.b_1 - b_2 - b_3 \in aff\ S$, since sum of coefficients is 1.

   $p_2 = 2.b_1 + b_3 \notin\ aff\ S$, since sum of coefficients is not 1.

   $p_3 = -b_1 + 2.b_2 + 0.b_3 \in aff\ S$, since sum of coefficients is 1.

   b) $p_1 = -4.b_1 + 2.b_2 + 3.b_3 \in aff\ S$, since sum of coefficients is 1.

   $p_2 = \frac{1}{5}.b_1 + \frac{1}{2}.b_2 + \frac{3}{10}.b_3 \in aff\ S$, since sum of coefficients is 1.

   $p_3 = b_1 + b_2 + b_3 \notin aff\ S$, since sum of coefficients is not 1.

3. $v_1 = \begin{bmatrix} 1 \\ -3 \\ 4 \end{bmatrix}, v_2 = \begin{bmatrix} 4 \\ 4 \\ -6 \end{bmatrix}$ other answers possible

4. a) true   b) false   c) true   d) false   e) true

5. a) $p_1 = 2.v_1 + v_2 - v_3,\quad p_1 \in Span\ S$ but $p_1 \notin aff\ S$

   $p_2 = 2.v_1 - 4.v_2 + 3.v_3,\quad p_2 \in Span\ S$ and $p_2 \in aff\ S$

   $p_3 \notin Span\ S$ hence $p_3 \notin aff\ S$

b) $p_1 = 3.v_1 - v_2 + v_3, \quad p_1 \in Span\ S$ but $p_1 \notin aff\ S$

$\quad p_2 \notin Span\ S$ hence $p_2 \notin aff\ S$

$\quad p_3 = -2.v_1 + 6.v_2 - 3.v_3, \quad p_3 \in Span\ S$ and $p_3 \in aff\ S$

6. a) Affinely dependent and $2.v_1 + v_2 - 3.v_3 = 0$

   b) Affinely independent

   c) affinely independent $v_4 = 16 * v_1 + 5 * v_2 - 3 * v_3$ but sum of weights in linear combimation is not zero.

   d) affinely dependent $-6.v_1 + 3.v_2 - 2.v_3 + 5.v_4 = 0$

   e) affinely dependent $-4.v_1 + 5.v_2 - 4.v_3 + 3.v_4 = 0$

   f) affinely independent

7. $c_1 v_1 + c_2 v_3 + c_3 v_3 = 0, c_1 + c_2 + c_3 = 0$ implies that $c_1 = c_2 = c_3 = 0$

8. $4 * v_1 - 5 * v_2 - 3 * v_3 + 4 * v_4 = 0$ sum of weights is zero.

9. $p_1 = 0.v_1 + \frac{1}{3}v_2 + \frac{2}{3}v_3, p_1 \in convS$

   $p_2 = \frac{13}{24}.v_1 + \frac{1}{36}v_2 + \frac{2}{9}v_3, p_2 \notin convS$

10. $p_1 = \frac{-1}{6}.v_1 + \frac{1}{3}v_2 + \frac{2}{3}v_3 + \frac{1}{6}.v_4, p_1 \notin convS$

    $p_2 = \frac{1}{3}.v_1 + \frac{1}{3}v_2 + \frac{1}{6}v_3 + \frac{1}{6}.v_4, p_2 \in convS$

11. $p_1 = \frac{1}{2}v_1 + \frac{-1}{2}v_2 + v_3$ hence $p_1 \notin conv\ S$

    $p_2 = \frac{1}{4}v_1 + \frac{1}{4}v_2 + \frac{1}{2}v_3$ hence $p_2 \in conv\ S$

12. Hint: Draw the curve $y = \frac{1}{x}$, $x > 0$ and justify your answer

13. Hint:Use scilab to sketch S and justify

14. Let $P = \begin{bmatrix} a \\ b \end{bmatrix}$ be an arbitrary point in S.

    a) P is on line segment through 0 and $\begin{bmatrix} \frac{b}{a} \\ \frac{a}{b} \end{bmatrix}$

    b) P is on line segment through 0 and $\begin{bmatrix} a \\ sina \end{bmatrix}$

    c) P is on line segment through 0 and $\begin{bmatrix} a \\ \sqrt{a} \end{bmatrix}$

15. None are in conv S

16. None are in conv S